HOLT SCIENCE & TECHNOLOGY

Sound and Light

HOLT, RINEHART AND WINSTON

A Harcourt Education Company

Orlando • **Austin** • New York • San Diego • Toronto • London

Acknowledgments

Contributing Authors

Leila Dumas, MA
Former Physics Teacher
Lago Vista, Texas

Inclusion and Special Needs Consultant

Ellen McPeek Glisan
Special Needs Consultant
San Antonio, Texas

Safety Reviewer

Jack Gerlovich, Ph.D.
Associate Professor
School of Education
Drake University
Des Moines, Iowa

Academic Reviewers

Howard L. Brooks, Ph.D.
Professor of Physics & Astronomy
Department of Physics & Astronomy
DePauw University
Greencastle, Indiana

Simonetta Frittelli, Ph.D.
Associate Professor
Department of Physics
Duquesne University
Pittsburgh, Pennsylvania

David S. Hall, Ph.D.
Assistant Professor of Physics
Department of Physics
Amherst College
Amherst, Massachusetts

William H. Ingham, Ph.D.
Professor of Physics
James Madison University
Harrisonburg, Virginia

David Lamp, Ph.D.
Associate Professor of Physics
Physics Department
Texas Tech University
Lubbock, Texas

Mark Mattson, Ph.D.
Director, College of Science and Mathematics Learning Center
James Madison University
Harrisonburg, Virginia

Richard F. Niedziela, Ph.D.
Assistant Professor of Chemistry
Department of Chemistry
DePaul University
Chicago, Illinois

H. Michael Sommermann, Ph.D.
Professor of Physics
Physics Department
Westmont College
Santa Barbara, California

Lab Testing

Barry L. Bishop
Science Teacher and Department Chair
San Rafael Junior High School
Ferron, Utah

Paul Boyle
Science Teacher
Perry Heights Middle School
Evansville, Indiana

Jennifer Ford
Science Teacher and Dept. Chair
North Ridge Middle School
North Richland Hills, Texas

Tracy Jahn
Science Teacher
Berkshire Junior-Senior High School
Canaan, New York

Edith C. McAlanis
Science Teacher and Department Chair
Socorro Middle School
El Paso, Texas

Kevin McCurdy, Ph.D.
Science Teacher
Elmwood Junior High School
Rogers, Arkansas

Terry J. Rakes
Science Teacher
Elmwood Junior High School
Rogers, Arkansas

Patricia McFarlane Soto
Science Teacher and Department Chair
G. W. Carver Middle School
Miami, Florida

Printed in the United States of America

ISBN 978-0-03-050132-6
ISBN 0-03-050132-6
9 10 11 12 0868 16 15 14 13
4500409847

0 Sound and Light

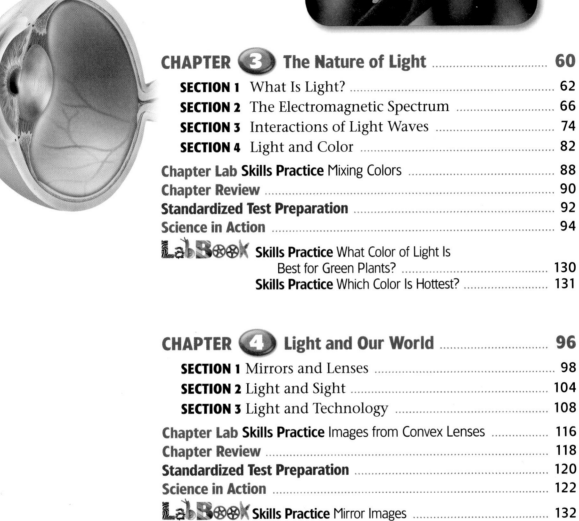

Labs and Activities

How to Use Your Textbook

Your Roadmap for Success with Holt Science and Technology

What You Will Learn

At the beginning of every section you will find the section's objectives and vocabulary terms. The objectives tell you what you'll need to know after you finish reading the section.

Vocabulary terms are listed for each section. Learn the definitions of these terms because you will most likely be tested on them. Each term is highlighted in the text and is defined at point of use and in the margin. You can also use the glossary to locate definitions quickly.

STUDY TIP Reread the objectives and the definitions to the terms when studying for a test to be sure you know the material.

Get Organized

A Reading Strategy at the beginning of every section provides tips to help you organize and remember the information covered in the section. Keep a science notebook so that you are ready to take notes when your teacher reviews the material in class. Keep your assignments in this notebook so that you can review them when studying for the chapter test.

SECTION 3

Interactions of Sound Waves

Have you ever heard of a sea canary? It's not a bird! It's a whale! Beluga whales are sometimes called sea canaries because of the many different sounds they make.

Dolphins, beluga whales, and many other animals that live in the sea use sound to communicate. Beluga whales also rely on reflected sound waves to find fish, crabs, and shrimp to eat. In this section, you will learn about reflection and other interactions of sound waves. You will also learn how bats, dolphins, and whales use sound to find food.

What You Will Learn
- Explain how echoes are made, and describe their use in locating objects.
- List examples of constructive and destructive interference of sound waves.
- Explain what resonance is.

Vocabulary
echo   sonic boom
echolocation standing wave
interference resonance

READING STRATEGY

Paired Summarizing Read this section silently. In pairs, take turns summarizing the material. Stop to discuss ideas that seem confusing.

Reflection of Sound Waves

Reflection is the bouncing back of a wave after it strikes a barrier. You're probably already familiar with a reflected sound wave, otherwise known as an **echo**. The strength of a reflected sound wave depends on the reflecting surface. Sound waves reflect best off smooth, hard surfaces. Look at **Figure 1**. A shout in an empty gymnasium can produce an echo, but a shout in an auditorium usually does not.

The difference is that the walls of an auditorium are usually designed so that they absorb sound. If sound waves hit a flat, hard surface, they will reflect back. Reflection of sound waves doesn't matter much in a gymnasium. But you don't want to hear echoes while listening to a musical performance!

echo a reflected sound wave

Figure 1   Sound Reflection and Absorption

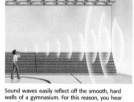

Sound waves easily reflect off the smooth, hard walls of a gymnasium. For this reason, you hear an echo.

In well-designed auditoriums, echoes are reduced by soft materials that absorb sound waves and by irregular shapes that scatter sound waves.

612   Chapter 21   The Nature of Sound

↗ Be Resourceful—Use the Web

SciLinks boxes in your textbook take you to resources that you can use for science projects, reports, and research papers. Go to **scilinks.org** and type in the **SciLinks code** to find information on a topic.

Visit go.hrw.com
Check out the **Current Science®** magazine articles and other materials that go with your textbook at **go.hrw.com.** Click on the textbook icon and the table of contents to see all of the resources for each chapter.

Use the Illustrations and Photos

Art shows complex ideas and processes. Learn to analyze the art so that you better understand the material you read in the text.

Tables and graphs display important information in an organized way to help you see relationships.

A picture is worth a thousand words. Look at the photographs to see relevant examples of science concepts that you are reading about.

Answer the Section Reviews

Section Reviews test your knowledge of the main points of the section. Critical Thinking items challenge you to think about the material in greater depth and to find connections that you infer from the text.

STUDY TIP When you can't answer a question, reread the section. The answer is usually there.

Do Your Homework

Your teacher may assign worksheets to help you understand and remember the material in the chapter.

STUDY TIP Don't try to answer the questions without reading the text and reviewing your class notes. A little preparation up front will make your homework assignments a lot easier. Answering the items in the Chapter Review will help prepare you for the chapter test.

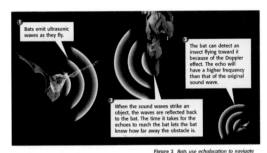

Bats emit ultrasonic waves as they fly.

The bat can detect an insect flying toward it because of the Doppler effect. The echo will have a higher frequency than that of the original sound wave.

When the sound waves strike an object, the waves are reflected back to the bat. The time it takes for the echoes to reach the bat lets the bat know how far away the obstacle is.

Figure 2 Bats use echolocation to navigate around barriers and to find insects to eat.

Echolocation

Beluga whales use echoes to find food. The use of reflected sound waves to find objects is called **echolocation.** Other animals—such as dolphins, bats, and some kinds of birds—also use echolocation to hunt food and find objects in their paths. **Figure 2** shows how echolocation works. Animals that use echolocation can tell how far away something is based on how long it takes sound waves to echo back to their ears. Some animals, such as bats, also make use of the Doppler effect to tell if another moving object, such as an insect, is moving toward it or away from it.

echolocation the process of using reflected sound waves to find objects; used by animals such as bats

Resonance in Musical Instruments

Musical instruments use resonance to make sound. In wind instruments, vibrations are caused by blowing air into the mouthpiece. The vibrations make a sound, which is amplified when it forms a standing wave inside the instrument.

String instruments also resonate when they are played. An acoustic guitar, such as the one shown in **Figure 10**, has a hollow body. When the strings vibrate, sound waves enter the body of the guitar. Standing waves form inside the body of the guitar, and the sound is amplified.

Figure 10 The body of a guitar resonates when the guitar is strummed.

SECTION Review

Summary

- Echoes are reflected sound waves.
- Some animals can use echolocation to find food or to navigate around objects.
- People use echolocation technology in many underwater applications.
- Ultrasonography uses sound reflection for medical applications.
- Sound barriers and shock waves are created by interference.
- Standing waves form at an object's resonant frequencies.
- Resonance happens when a vibrating object causes a second object to vibrate at one of its resonant frequencies.

Using Key Terms

1. Use the following terms in the same sentence: *echo* and *echolocation*.

Complete each of the following sentences by choosing the correct term from the word bank.

interference   standing wave
sonic boom   resonance

2. When you pluck a string on a musical instrument, a(n) _____ forms.

3. When a vibrating object causes a nearby object to vibrate, _____ results.

Understanding Key Ideas

4. What causes an echo?
   a. reflection
   b. resonance
   c. constructive interference
   d. destructive interference

5. Describe a place in which you would expect to hear echoes.

6. How do bats use echoes to find insects to eat?

7. Give one example each of constructive and destructive interference of sound waves.

Math Skills

8. Sound travels through air at 343 m/s at 20°C. A bat emits an ultrasonic squeak and hears the echo 0.05 s later. How far away was the object that reflected it? (Hint: Remember that the sound must travel *to* the object and *back* to the bat.)

Critical Thinking

9. **Applying Concepts** Your friend is playing a song on a piano. Whenever your friend hits a certain key, the lamp on top of the piano rattles. Explain why the lamp rattles.

10. **Making Comparisons** Compare sonar and ultrasonography in locating objects.

SciLINKS.
Developed and maintained by the National Science Teachers Association

For a variety of links related to this chapter, go to www.scilinks.org

Topic: Interactions of Sound Waves
SciLinks code: HSM0804

617

Visit Holt Online Learning

If your teacher gives you a special password to log onto the **Holt Online Learning** site, you'll find your complete textbook on the Web. In addition, you'll find some great learning tools and practice quizzes. You'll be able to see how well you know the material from your textbook.

SAFETY FIRST!

Exploring, inventing, and investigating are essential to the study of science. However, these activities can also be dangerous. To make sure that your experiments and explorations are safe, you must be aware of a variety of safety guidelines. You have probably heard of the saying, "It is better to be safe than sorry." This is particularly true in a science classroom where experiments and explorations are being performed. Being uninformed and careless can result in serious injuries. Don't take chances with your own safety or with anyone else's.

The following pages describe important guidelines for staying safe in the science classroom. Your teacher may also have safety guidelines and tips that are specific to your classroom and laboratory. Take the time to be safe.

Safety Rules!

Start Out Right

Always get your teacher's permission before attempting any laboratory exploration. Read the procedures carefully, and pay particular attention to safety information and caution statements. If you are unsure about what a safety symbol means, look it up or ask your teacher. You cannot be too careful when it comes to safety. If an accident does occur, inform your teacher immediately regardless of how minor you think the accident is.

If you are instructed to note the odor of a substance, wave the fumes toward your nose with your hand. Never put your nose close to the source.

Safety Symbols

All of the experiments and investigations in this book and their related worksheets include important safety symbols to alert you to particular safety concerns. Become familiar with these symbols so that when you see them, you will know what they mean and what to do. It is important that you read this entire safety section to learn about specific dangers in the laboratory.

Eye protection

Clothing protection

Hand safety

Heating safety

Electric safety

Chemical safety

Animal safety

Sharp object

Plant safety

Eye Safety

Wear safety goggles when working around chemicals, acids, bases, or any type of flame or heating device. Wear safety goggles any time there is even the slightest chance that harm could come to your eyes. If any substance gets into your eyes, notify your teacher immediately and flush your eyes with running water for at least 15 minutes. Treat any unknown chemical as if it were a dangerous chemical. Never look directly into the sun. Doing so could cause permanent blindness.

Avoid wearing contact lenses in a laboratory situation. Even if you are wearing safety goggles, chemicals can get between the contact lenses and your eyes. If your doctor requires that you wear contact lenses instead of glasses, wear eye-cup safety goggles in the lab.

Safety Equipment

Know the locations of the nearest fire alarms and any other safety equipment, such as fire blankets and eyewash fountains, as identified by your teacher, and know the procedures for using the equipment.

Neatness

Keep your work area free of all unnecessary books and papers. Tie back long hair, and secure loose sleeves or other loose articles of clothing, such as ties and bows. Remove dangling jewelry. Don't wear open-toed shoes or sandals in the laboratory. Never eat, drink, or apply cosmetics in a laboratory setting. Food, drink, and cosmetics can easily become contaminated with dangerous materials.

Certain hair products (such as aerosol hair spray) are flammable and should not be worn while working near an open flame. Avoid wearing hair spray or hair gel on lab days.

Sharp/Pointed Objects

Use knives and other sharp instruments with extreme care. Never cut objects while holding them in your hands. Place objects on a suitable work surface for cutting.

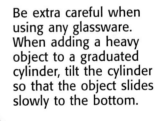

Be extra careful when using any glassware. When adding a heavy object to a graduated cylinder, tilt the cylinder so that the object slides slowly to the bottom.

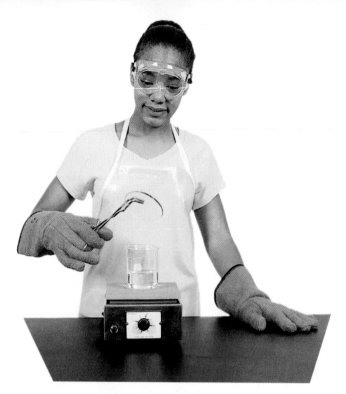

Chemicals

Wear safety goggles when handling any potentially dangerous chemicals, acids, or bases. If a chemical is unknown, handle it as you would a dangerous chemical. Wear an apron and protective gloves when you work with acids or bases or whenever you are told to do so. If a spill gets on your skin or clothing, rinse it off immediately with water for at least 5 minutes while calling to your teacher.

Never mix chemicals unless your teacher tells you to do so. Never taste, touch, or smell chemicals unless you are specifically directed to do so. Before working with a flammable liquid or gas, check for the presence of any source of flame, spark, or heat.

Heat

Wear safety goggles when using a heating device or a flame. Whenever possible, use an electric hot plate as a heat source instead of using an open flame. When heating materials in a test tube, always angle the test tube away from yourself and others. To avoid burns, wear heat-resistant gloves whenever instructed to do so.

Electricity

Be careful with electrical cords. When using a microscope with a lamp, do not place the cord where it could trip someone. Do not let cords hang over a table edge in a way that could cause equipment to fall if the cord is accidentally pulled. Do not use equipment with damaged cords. Be sure that your hands are dry and that the electrical equipment is in the "off" position before plugging it in. Turn off and unplug electrical equipment when you are finished.

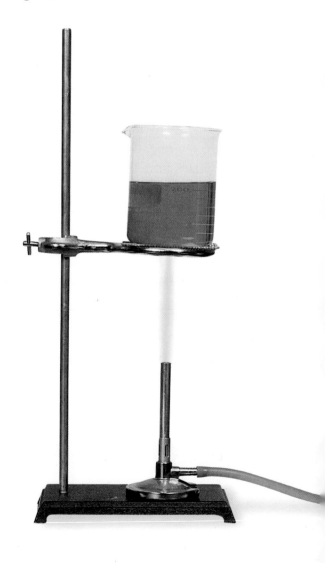

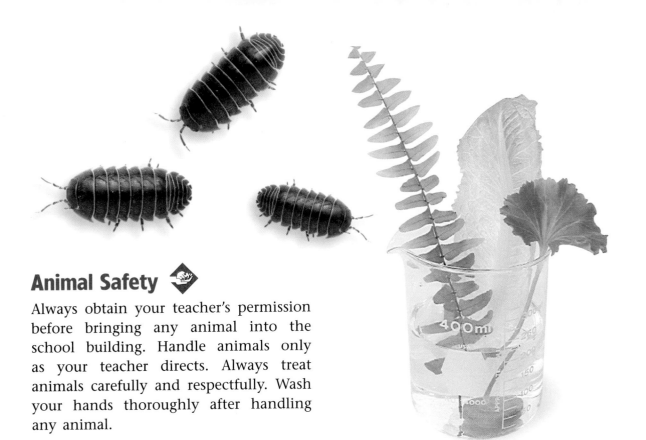

Animal Safety

Always obtain your teacher's permission before bringing any animal into the school building. Handle animals only as your teacher directs. Always treat animals carefully and respectfully. Wash your hands thoroughly after handling any animal.

Plant Safety

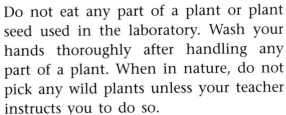

Do not eat any part of a plant or plant seed used in the laboratory. Wash your hands thoroughly after handling any part of a plant. When in nature, do not pick any wild plants unless your teacher instructs you to do so.

Glassware

Examine all glassware before use. Be sure that glassware is clean and free of chips and cracks. Report damaged glassware to your teacher. Glass containers used for heating should be made of heat-resistant glass.

1

The Energy of Waves

The Big Idea

Waves transfer energy, have describable properties, and interact in predictable ways.

About the PHOTO

A surfer takes advantage of a wave's energy to catch an exciting ride. The ocean wave that this surfer is riding is just one type of wave. You are probably familiar with water waves. But did you know that light, sound, and even earthquakes are waves? From music to television, waves play an important role in your life every day.

PRE-READING ACTIVITY

FOLDNOTES

Three-Panel Flip Chart
Before you read the chapter, create the FoldNote entitled "Three-Panel Flip Chart" described in the **Study Skills** section of the Appendix. Label the flaps of the three-panel flip chart with "The nature of waves," "Properties of waves," and "Wave interactions." As you read the chapter, write information you learn about each category under the appropriate flap.

START-UP ACTIVITY

Energetic Waves

In this activity, you will observe the movement of a wave. Then, you will determine the source of the wave's energy.

Procedure

1. Tie one end of a **piece of rope** to the back of a **chair.**

2. Hold the other end in one hand, and stand away from the chair so that the rope is almost straight but is not pulled tight.

3. Move the rope up and down quickly to create a wave. Repeat this step several times. Record your observations.

Analysis

1. In which direction does the wave move?

2. How does the movement of the rope compare with the movement of the wave?

3. Where does the energy of the wave come from?

The Nature of Waves

Imagine that your family has just returned home from a day at the beach. You had fun playing in the ocean under a hot sun. You put some cold pizza in the microwave for dinner, and you turn on the radio. Just then, the phone rings. It's your friend calling to ask about homework.

In the events described above, how many different waves were present? Believe it or not, there were at least five! Can you name them? Here's a hint: A **wave** is any disturbance that transmits energy through matter or empty space. Okay, here are the answers: water waves in the ocean; light waves from the sun; microwaves inside the microwave oven; radio waves transmitted to the radio; and sound waves from the radio, telephone, and voices. Don't worry if you didn't get very many. You will be able to name them all after you read this section.

✔️ **Reading Check** What do all waves have in common? (*See the Appendix for answers to Reading Checks.*)

Wave Energy

Energy can be carried away from its source by a wave. You can observe an example of a wave if you drop a rock in a pond. Waves from the rock's splash carry energy away from the splash. However, the material through which the wave travels does not move with the energy. Look at **Figure 1.** Can you move a leaf on a pond if you are standing on the shore? You can make the leaf bob up and down by making waves that carry enough energy through the water. But you would not make the leaf move in the same direction as the wave.

What You Will Learn

- Describe how waves transfer energy without transferring matter.
- Distinguish between waves that require a medium and waves that do not.
- Explain the difference between transverse and longitudinal waves.

Vocabulary

wave
medium

transverse wave
longitudinal wave

Discussion Read this section silently. Write down questions that you have about this section. Discuss your questions in a small group.

wave a periodic disturbance in a solid, liquid, or gas as energy is transmitted through a medium

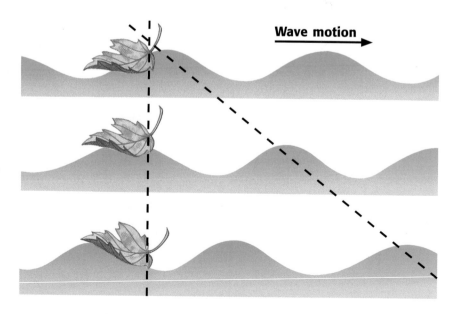

Wave motion

Figure 1 *Waves on a pond move toward the shore, but the water and the leaf floating on the surface only bob up and down.*

Waves and Work

As a wave travels, it does work on everything in its path. The waves in a pond do work on the water to make it move up and down. The waves also do work on anything floating on the water's surface. For example, boats and ducks bob up and down with waves. The fact that these objects move tells you that the waves are transferring energy.

Energy Transfer Through a Medium

Most waves transfer energy by the vibration of particles in a medium. A **medium** is a substance through which a wave can travel. A medium can be a solid, a liquid, or a gas. The plural of *medium* is *media*.

When a particle vibrates (moves back and forth, as in **Figure 2**), it can pass its energy to a particle next to it. The second particle will vibrate like the first particle does. In this way, energy is transmitted through a medium.

Sound waves need a medium. Sound energy travels by the vibration of particles in liquids, solids, and gases. If there are no particles to vibrate, no sound is possible. If you put an alarm clock inside a jar and remove all the air from the jar to create a vacuum, you will not be able to hear the alarm.

Other waves that need a medium include ocean waves, which move through water, and waves that are carried on guitar and cello strings when they vibrate. Waves that need a medium are called *mechanical waves*. **Figure 3** shows the effect of a mechanical wave in Earth's crust: an earthquake.

Figure 2 *A vibration is one complete back-and-forth motion of an object.*

medium a physical environment in which phenomena occur

Figure 3 *Earthquakes cause seismic waves to travel through Earth's crust. The energy they carry can be very destructive to anything on the ground.*

The Nature of Waves **5**

Figure 4 *Light waves are electromagnetic waves, which do not need a medium. Light waves from the Crab nebula, shown here, travel through the vacuum of space billions of miles to Earth, where they can be detected with a telescope.*

Energy Transfer Without a Medium

Some waves can transfer energy without going through a medium. Visible light is one example. Other examples include microwaves made by microwave ovens, TV and radio signals, and X rays used by dentists and doctors. These waves are *electromagnetic waves.*

Although electromagnetic waves do not need a medium, they can go through matter, such as air, water, and glass. The energy that reaches Earth from the sun comes through electromagnetic waves, which go through space. As shown in **Figure 4,** you can see light from stars because electromagnetic waves travel through space to Earth. Light is an electromagnetic wave that your eyes can see.

✓ Reading Check How do electromagnetic waves differ from mechanical waves?

CONNECTION TO Astronomy

Light Speed Light waves from stars and galaxies travel great distances that are best expressed in light-years. A light-year is the distance a ray of light can travel in one year. Some of the light waves from these stars have traveled billions of light-years before reaching Earth. Do the following calculation in your **science journal:** If light travels at a speed of 300,000,000 m/s, what distance is a light-minute? (Hint: There are 60 s in a minute.)

ACTIVITY

Types of Waves

All waves transfer energy by repeated vibrations. However, waves can differ in many ways. Waves can be classified based on the direction in which the particles of the medium vibrate compared with the direction in which the waves move. The two main types of waves are *transverse waves* and *longitudinal* (LAHN juh TOOD'n uhl) *waves*. Sometimes, a transverse wave and a longitudinal wave can combine to form another kind of wave called a *surface wave*.

Transverse Waves

Waves in which the particles vibrate in an up-and-down motion are called **transverse waves.** *Transverse* means "moving across." The particles in this kind of wave move across, or perpendicularly to, the direction that the wave is going. To be *perpendicular* means to be "at right angles."

A wave moving on a rope is an example of a transverse wave. In **Figure 5,** you can see that the points along the rope vibrate perpendicularly to the direction the wave is going. The highest point of a transverse wave is called a *crest*, and the lowest point between each crest is called a *trough* (TRAWF). Although electromagnetic waves do not travel by vibrating particles in a medium, all electromagnetic waves are considered transverse waves. The reason is that the waves are made of vibrations that are perpendicular to the direction of motion.

INTERNET ACTIVITY

For another activity related to this chapter, go to **go.hrw.com** and type in the keyword **HP5WAVW.**

transverse wave a wave in which the particles of the medium move perpendicularly to the direction the wave is traveling

Figure 5   Motion of a Transverse Wave

A wave on a rope is a transverse wave because the particles of the medium vibrate perpendicularly to the direction the wave moves.

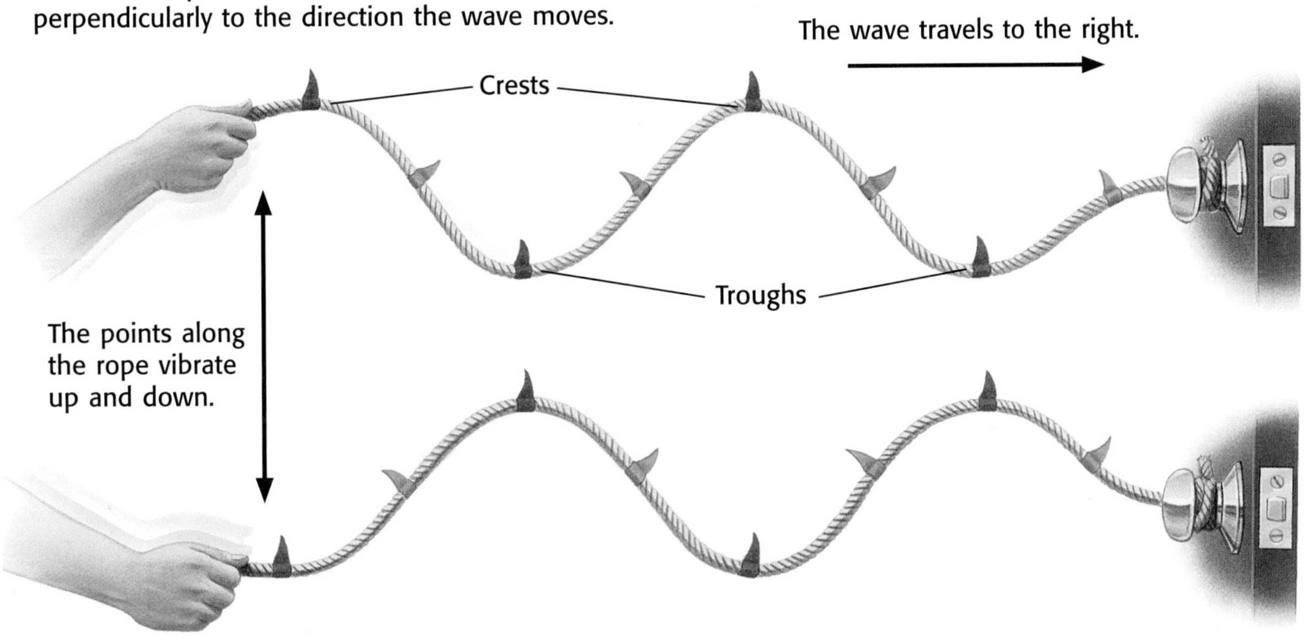

The wave travels to the right.

Crests

Troughs

The points along the rope vibrate up and down.

Figure 6 **Comparing Longitudinal and Transverse Waves**

Pushing a spring back and forth creates a longitudinal wave, much the same way that shaking a rope up and down creates a transverse wave.

Rarefactions   Compressions

Longitudinal wave

Troughs   Crests

Transverse wave

Longitudinal Waves

longitudinal wave a wave in which the particles of the medium vibrate parallel to the direction of wave motion

In a **longitudinal wave,** the particles of the medium vibrate back and forth along the path that the wave moves. You can make a longitudinal wave on a spring. When you push on the end of the spring, the coils of the spring crowd together. A part of a longitudinal wave where the particles are crowded together is called a *compression*. When you pull back on the end of the spring, the coils are pulled apart. A part where the particles are spread apart is a *rarefaction* (RER uh FAK shuhn). Compressions and rarefactions are like the crests and troughs of a transverse wave, as shown in **Figure 6.**

Sound Waves

A sound wave is an example of a longitudinal wave. Sound waves travel by compressions and rarefactions of air particles. **Figure 7** shows how a vibrating drum forms compressions and rarefactions in the air around it.

✓ *Reading Check* What kind of wave is a sound wave?

Figure 7 *Sound energy is carried away from a drum by a longitudinal wave through the air.*

When the drumhead moves out after being hit, a compression is created in the air particles.

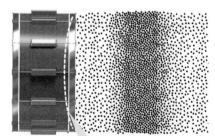

When the drumhead moves back in, a rarefaction is created.

Combinations of Waves

When waves form at or near the boundary between two media, a transverse wave and a longitudinal wave can combine to form a *surface wave*. An example is shown in **Figure 8.** Surface waves look like transverse waves, but the particles of the medium in a surface wave move in circles rather than up and down. The particles move forward at the crest of each wave and move backward at the trough.

Figure 8 *Ocean waves are surface waves. A floating bottle shows the circular motion of particles in a surface wave.*

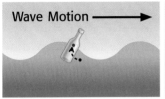
Wave Motion ⟶

SECTION Review

Summary

- A wave is a disturbance that transmits energy.
- The particles of a medium do not travel with the wave.
- Mechanical waves require a medium, but electromagnetic waves do not.
- Particles in a transverse wave vibrate perpendicularly to the direction the wave travels.
- Particles in a longitudinal wave vibrate parallel to the direction that the wave travels.

Using Key Terms

Complete each of the following sentences by choosing the correct term from the word bank.

transverse wave	wave
longitudinal wave	medium

1. In a ___, the particles vibrate parallel to the direction that the wave travels.

2. Mechanical waves require a ___ through which to travel.

3. Any ___ transmits energy through vibrations.

4. In a ___, the particles vibrate perpendicularly to the direction that the wave travels.

Understanding Key Ideas

5. Waves transfer
   - **a.** matter.   **c.** particles.
   - **b.** energy.   **d.** water.

6. Name a kind of wave that does not require a medium.

Critical Thinking

7. **Applying Concepts** Sometimes, people at a sports event do "the wave." Is this a real example of a wave? Why or why not?

8. **Making Inferences** Why can supernova explosions in space be seen but not heard on Earth?

Interpreting Graphics

9. Look at the figure below. Which part of the wave is the crest? Which part of the wave is the trough?

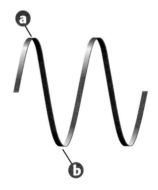

For a variety of links related to this chapter, go to www.scilinks.org
Topic: The Nature of Waves; Types of Waves
SciLinks code: HSM1017; HSM1574

Developed and maintained by the National Science Teachers Association

Properties of Waves

You are in a swimming pool, floating on your air mattress, enjoying a gentle breeze. Your friend does a "cannonball" from the high dive nearby. Suddenly, your mattress is rocking wildly on the waves generated by the huge splash.

The breeze generates waves in the water as well, but they are very different from the waves created by your diving friend. The waves made by the breeze are shallow and close together, while the waves from your friend's splash are tall and widely spaced. Properties of waves, such as the height of the waves and the distance between crests, are useful for comparing and describing waves.

Amplitude

If you tie one end of a rope to the back of a chair, you can create waves by moving the free end up and down. If you shake the rope a little, you will make a shallow wave. If you shake the rope hard, you will make a tall wave.

The **amplitude** of a wave is related to its height. A wave's amplitude is the maximum distance that the particles of a medium vibrate from their rest position. The rest position is the point where the particles of a medium stay when there are no disturbances. The larger the amplitude is, the taller the wave is. **Figure 1** shows how the amplitude of a transverse wave may be measured.

Larger Amplitude—More Energy

When using a rope to make waves, you have to work harder to create a wave with a large amplitude than to create one with a small amplitude. The reason is that it takes more energy to move the rope farther from its rest position. Therefore, a wave with a large amplitude carries more energy than a wave with a small amplitude does.

What You Will Learn

- Identify and describe four wave properties.
- Explain how frequency and wavelength are related to the speed of a wave.

Vocabulary

amplitude   frequency
wavelength   wave speed

READING STRATEGY

Mnemonics As you read this section, create a mnemonic device to help you remember the wave equation.

amplitude the maximum distance that the particles of a wave's medium vibrate from their rest position

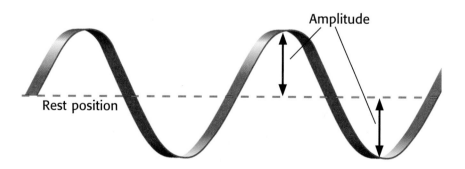

Figure 1 *The amplitude of a transverse wave is measured from the rest position to the crest or to the trough of the wave.*

Amplitude

Rest position

Wavelength

Another property of waves is wavelength. A **wavelength** is the distance between any two crests or compressions next to each other in a wave. The distance between two troughs or rarefactions next to each other is also a wavelength. In fact, the wavelength can be measured from any point on a wave to the next corresponding point on the wave. Wavelength is measured the same way in both a longitudinal wave and a transverse wave, as shown in **Figure 2.**

Shorter Wavelength—More Energy

If you are making waves on either a spring or a rope, the rate at which you shake it will determine whether the wavelength is short or long. If you shake it rapidly back and forth, the wavelength will be shorter. If you are shaking it rapidly, you are putting more energy into it than if you were shaking it more slowly. So, a wave with a shorter wavelength carries more energy than a wave with a longer wavelength does.

✓ **Reading Check** How does shaking a rope at different rates affect the wavelength of the wave that moves through the rope? (*See the Appendix for answers to Reading Checks.*)

wavelength the distance from any point on a wave to an identical point on the next wave

Springy Waves

1. Hold a coiled **spring toy** on the floor between you and a classmate so that the spring is straight. This is the rest position.

2. Move one end of the spring back and forth at a constant rate. Note the wavelength of the wave you create.

3. Increase the amplitude of the waves. What did you have to do? How did the change in amplitude affect the wavelength?

4. Now, shake the spring back and forth about twice as fast as you did before. What happens to the wavelength? Record your observations.

Figure 2   Measuring Wavelengths

Wavelength can be measured from any two corresponding points that are adjacent on a wave.

Longitudinal wave

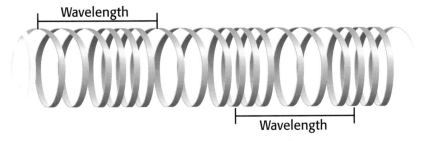

Transverse wave

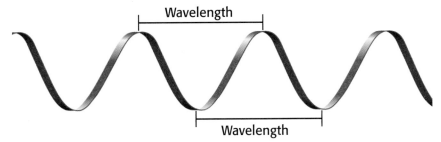

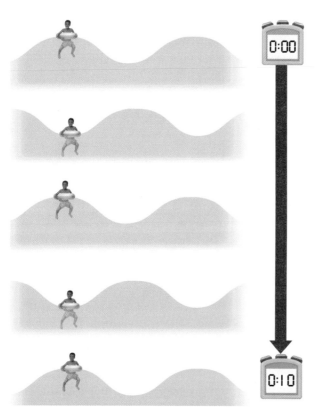

Figure 3 *Frequency can be measured by counting how many waves pass by in a certain amount of time. Here, two waves went by in 10 s, so the frequency is 2/10 s = 0.2 Hz.*

Frequency

Think about making rope waves again. The number of waves that you can make in 1 s depends on how quickly you move the rope. If you move the rope slowly, you make only a small number of waves each second. If you move it quickly, you make a large number of waves. The number of waves produced in a given amount of time is the **frequency** of the wave. Frequency is usually expressed in *hertz* (Hz). For waves, one hertz equals one wave per second (1 Hz = 1/s). **Figure 3** shows a wave with a frequency of 0.2 Hz.

✔ **Reading Check** If you make three rope waves per second, what is the frequency of the wave?

Higher Frequency—More Energy

To make high-frequency waves in a rope, you must shake the rope quickly back and forth. To shake a rope quickly takes more energy than to shake it slowly. Therefore, if the amplitudes are equal, high-frequency waves carry more energy than low-frequency waves.

Wave Speed

Wave speed is the speed at which a wave travels. Wave speed (*v*) can be calculated using wavelength (*λ*, the Greek letter *lambda*) and frequency (*f*), by using the *wave equation*, which is shown below:

$$v = \lambda \times f$$

Wave Calculations Determine the wave speed of a wave that has a wavelength of 5 m and a frequency of 4 Hz.

Step 1: Write the equation for wave speed.

$$v = \lambda \times f$$

Step 2: Replace the *λ* and *f* with the values given in the problem, and solve.

$$v = 5 \text{ m} \times 4 \text{ Hz} = 20 \text{ m/s}$$

The equation for wave speed can also be rearranged to determine wavelength or frequency, as shown at top right.

$\lambda = \dfrac{v}{f}$ (Rearranged by dividing by *f*.)

$f = \dfrac{v}{\lambda}$ (Rearranged by dividing by *λ*.)

Now It's Your Turn

1. What is the frequency of a wave if the wave has a speed of 12 cm/s and a wavelength of 3 cm?

2. A wave has a frequency of 5 Hz and a wave speed of 18 m/s. What is its wavelength?

Frequency and Wavelength Relationship

Three of the basic properties of a wave are related to one another in the wave equation—wave speed, frequency, and wavelength. If you know any two of these properties of a wave, you can use the wave equation to find the third.

One of the things the wave equation tells you is the relationship between frequency and wavelength. If a wave is traveling a certain speed and you double its frequency, its wavelength will be cut in half. Or if you were to cut its frequency in half, the wavelength would be double what it was before. So, you can say that frequency and wavelength are *inversely* related. Think of a sound wave, traveling underwater at 1,440 m/s, given off by the sonar of a submarine like the one shown in **Figure 4.** If the sound wave has a frequency of 360 Hz, it will have a wavelength of 4.0 m. If the sound wave has twice that frequency, the wavelength will be 2.0 m, half as big.

The wave speed of a wave in a certain medium is the same no matter what the wavelength is. So, the wavelength and frequency of a wave depend on the wave speed, not the other way around.

Figure 4 *Submarines use sonar, sound waves in water, to locate underwater objects.*

frequency the number of waves produced in a given amount of time

wave speed the speed at which a wave travels through a medium

SECTION Review

Summary

- Amplitude is the maximum distance the particles of a medium vibrate from their rest position.
- Wavelength is the distance between two adjacent corresponding parts of a wave.
- Frequency is the number of waves that pass a given point in a given amount of time.
- Wave speed can be calculated by multiplying the wave's wavelength by the frequency.

Using Key Terms

1. In your own words, write a definition for each of the following terms: *amplitude, frequency,* and *wavelength.*

Understanding Key Ideas

2. Which of the following results in more energy in a wave?
   a. a smaller wavelength
   b. a lower frequency
   c. a shallower amplitude
   d. a lower speed

3. Draw a transverse wave, and label how the amplitude and wavelength are measured.

Math Skills

4. What is the speed (v) of a wave that has a wavelength (λ) of 2 m and a frequency (f) of 6 Hz?

Critical Thinking

5. **Making Inferences** A wave has a low speed but a high frequency. What can you infer about its wavelength?

6. **Analyzing Processes** Two friends blow two whistles at the same time. The first whistle makes a sound whose frequency is twice that of the sound made by the other whistle. Which sound will reach you first?

SC*i*LINKS.

NSTA
Developed and maintained by the
National Science Teachers Association

For a variety of links related to this chapter, go to www.scilinks.org

Topic: Properties of Waves
SciLinks code: HSM1236

Wave Interactions

If you've ever seen a planet in the night sky, you may have had a hard time telling it apart from a star. Both planets and stars shine brightly, but the light waves that you see are from very different sources.

All stars, including the sun, produce light. But planets do not produce light. So, why do planets shine so brightly? The planets and the moon shine because light from the sun *reflects* off them. Without reflection, you would not be able to see the planets. Reflection is one of the wave interactions that you will learn about in this section.

Reflection

Reflection happens when a wave bounces back after hitting a barrier. All waves—including water, sound, and light waves—can be reflected. The reflection of water waves is shown in **Figure 1.** Light waves reflecting off an object allow you to see that object. For example, light waves from the sun are reflected when they strike the surface of the moon. These reflected waves allow us to enjoy moonlit nights. A reflected sound wave is called an *echo*.

Waves are not always reflected when they hit a barrier. If all light waves were reflected when they hit your eyeglasses, you would not be able to see anything! A wave is *transmitted* through a substance when it passes through the substance.

What You Will Learn

● Describe reflection, refraction, diffraction, and interference.
● Compare destructive interference with constructive interference.
● Describe resonance, and give examples.

Vocabulary

reflection	interference
refraction	standing wave
diffraction	resonance

READING STRATEGY

Reading Organizer As you read this section, make a concept map by using the terms above.

reflection the bouncing back of a ray of light, sound, or heat when the ray hits a surface that it does not go through

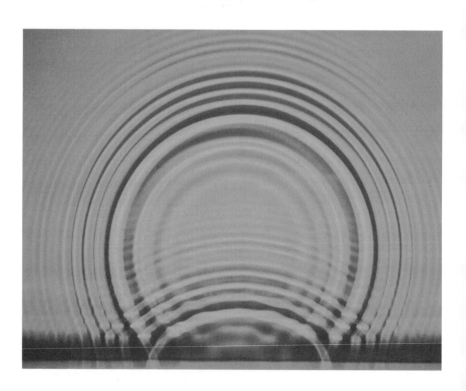

Figure 1 *These water waves are reflecting off the side of the container.*

Figure 2 *A light wave passing at an angle into a new medium— such as water—is refracted because the speed of the wave changes.*

Refraction

Try this simple activity: Place a pencil in a half-filled glass of water. Now, look at the pencil from the side. The pencil appears to be broken into two pieces! But as you can see when you take the pencil out of the water, it is still in one piece.

What you saw in this experiment was the result of the *refraction* of light waves. **Refraction** is the bending of a wave as the wave passes from one medium to another at an angle. Refraction of a flashlight beam as the beam passes from air to water is shown in **Figure 2.**

When a wave moves from one medium to another, the wave's speed changes. When a wave enters a new medium, the wave changes wavelength as well as speed. As a result, the wave bends and travels in a new direction.

refraction the bending of a wave as the wave passes between two substances in which the speed of the wave differs

☑ **Reading Check** What happens to a wave when it moves from one medium to another at an angle? (*See the Appendix for answers to Reading Checks.*)

Refraction of Different Colors

When light waves from the sun pass through a droplet of water in a cloud or through a prism, the light is refracted. But the different colors in sunlight are refracted by different amounts, so the light is *dispersed,* or spread out, into its separate colors. When sunlight is refracted this way through water droplets, you can see a rainbow. Why does that happen?

Although all light waves travel at the same speed through empty space, when light passes through a medium such as water or glass, the speed of the light wave depends on the wavelength of the light wave. Because the different colors of light have different wavelengths, their speeds are different, and they are refracted by different amounts. As a result, the colors are spread out, so you can see them individually.

CONNECTION TO Language Arts

WRITING SKILL **The Colors of the Rainbow** People have always been fascinated by the beautiful array of colors that results when sunlight strikes water droplets in the air to form a rainbow. The knowledge science gives us about how they form makes them no less breathtaking.

In the library, find a poem that you like about rainbows. In your **science journal,** copy the poem, and write a paragraph in which you discuss how your knowledge of refraction affects your opinion about the poem.

Figure 3 Diffraction Through an Opening

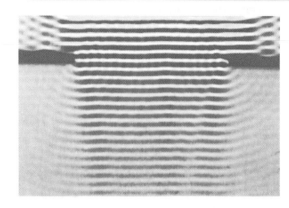

◀ If the barrier or opening is larger than the wavelength of the wave, there is only a small amount of diffraction.

◀ If the barrier or opening is the same size or smaller than the wavelength of an approaching wave, the amount of diffraction is large.

School to Home

What if Light Diffracted?

With an adult, take a walk around your neighborhood. Light waves diffract around corners of buildings much less than sound waves do. Imagine what would happen if light waves diffracted around corners much more than sound waves did. Write a paragraph in your **science journal** describing how this would change what you see and hear as you walk around your neighborhood.

Activity

diffraction a change in the direction of a wave when the wave finds an obstacle or an edge, such as an opening

Diffraction

Suppose you are walking down a city street and you hear music. The sound seems to be coming from around the corner, but you cannot see where the music is coming from because a building on the corner blocks your view. Why do sound waves travel around a corner better than light waves do?

Most of the time, waves travel in straight lines. For example, a beam of light from a flashlight is fairly straight. But in some circumstances, waves curve or bend when they reach the edge of an object. The bending of waves around a barrier or through an opening is known as **diffraction.**

If You Can Hear It, Why Can't You See It?

The amount of diffraction of a wave depends on its wavelength and the size of the barrier or opening the wave encounters, as shown in **Figure 3.** You can hear music around the corner of a building because sound waves have long wavelengths and are able to diffract around corners. However, you cannot see who is playing the music because the wavelengths of light waves are much shorter than sound waves, so light is not diffracted very much.

Interference

You know that all matter has volume. Therefore, objects cannot be in the same space at the same time. But waves are energy, not matter. So, more than one wave can be in the same place at the same time. In fact, two waves can meet, share the same space, and pass through each other! When two or more waves share the same space, they overlap. The result of two or more waves overlapping is called **interference. Figure 4** shows what happens when waves occupy the same space and interfere with each other.

interference the combination of two or more waves that results in a single wave

Constructive Interference

Constructive interference happens when the crests of one wave overlap the crests of another wave or waves. The troughs of the waves also overlap. When waves combine in this way, the energy carried by the waves is also able to combine. The result is a new wave that has higher crests and deeper troughs than the original waves had. In other words, the resulting wave has a larger amplitude than the original waves had.

✓ Reading Check How does constructive interference happen?

Figure 4 **Constructive and Destructive Interference**

Constructive Interference When waves combine by constructive interference, the combined wave has a larger amplitude.

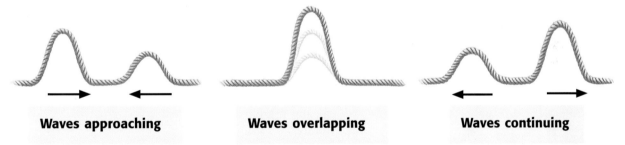

Waves approaching   **Waves overlapping**   **Waves continuing**

Destructive Interference When two waves with the same amplitude combine by destructive interference, they cancel each other out.

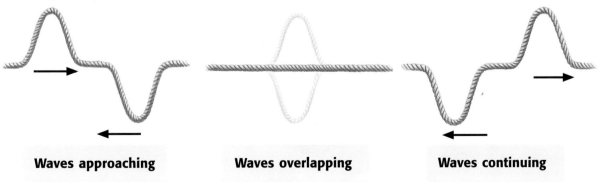

Waves approaching   **Waves overlapping**   **Waves continuing**

Destructive Interference

Destructive interference happens when the crests of one wave and the troughs of another wave overlap. The new wave has a smaller amplitude than the original waves had. When the waves involved in destructive interference have the same amplitude and meet each other at just the right time, the result is no wave at all.

Standing Waves

If you tie one end of a rope to the back of a chair and move the other end up and down, the waves you make go down the rope and are reflected back. If you move the rope at certain frequencies, the rope appears to vibrate in loops, as shown in **Figure 5.** The loops come from the interference between the wave you made and the reflected wave. The resulting wave is called a **standing wave.** In a standing wave, certain parts of the wave are always at the rest position because of total destructive interference between all the waves. Other parts have a large amplitude because of constructive interference.

A standing wave only *looks* as if it is standing still. Waves are actually going in both directions. Standing waves can be formed with transverse waves, such as when a musician plucks a guitar string, as well as with longitudinal waves.

Reading Check How can interference and reflection cause standing waves?

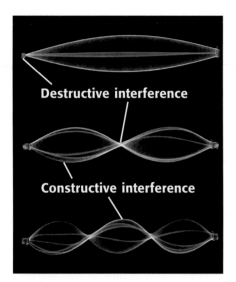

Figure 5 *When you move a rope at certain frequencies, you can create different standing waves.*

Figure 6 *A marimba produces notes through the resonance of air columns.*

a The marimba bars are struck with a mallet, causing the bars to vibrate.

b The vibrating bars cause the air in the columns to vibrate.

c The lengths of the columns have been adjusted so that the resonant frequency of the air column matches the frequency of the bar.

d The air column resonates with the bar, increasing the amplitude of the vibrations to produce a loud note.

Resonance

The frequencies at which standing waves are made are called *resonant frequencies*. When an object vibrating at or near the resonant frequency of a second object causes the second object to vibrate, **resonance** occurs. A resonating object absorbs energy from the vibrating object and vibrates, too. An example of resonance is shown in **Figure 6** on the previous page.

You may be familiar with another example of resonance at home—in your shower. When you sing in the shower, certain frequencies create standing waves in the air that fills the shower stall. The air resonates in much the same way that the air column in a marimba does. The amplitude of the sound waves becomes greater. So your voice sounds much louder.

standing wave a pattern of vibration that simulates a wave that is standing still

resonance a phenomenon that occurs when two objects naturally vibrate at the same frequency; the sound produced by one object causes the other object to vibrate

SECTION Review

Summary

- Waves reflect after hitting a barrier.
- Refraction is the bending of a wave when it passes through different media.
- Waves bend around barriers or through openings during diffraction.
- The result of two or more waves overlapping is called interference.
- Amplitude increases during constructive interference and decreases during destructive interference.
- Resonance occurs when a vibrating object causes another object to vibrate at one of its resonant frequencies.

Using Key Terms

Complete each of the following sentences by choosing the correct term from the word bank.

refraction	reflection
diffraction	interference

1. ___ happens when a wave passes from one medium to another at an angle.

2. The bending of a wave around a barrier is called ___.

3. We can see the moon because of the ___ of sunlight off it.

Understanding Key Ideas

4. The combining of waves as they overlap is known as
   a. interference.
   b. diffraction.
   c. refraction.
   d. resonance.

5. Name two wave interactions that can occur when a wave encounters a barrier.

6. Explain why you can hear two people talking even after they walk around a corner.

7. Explain what happens when two waves encounter one another in destructive interference.

Critical Thinking

8. **Making Inferences** Sometimes, when music is played loudly, you can feel your body shake. Explain what is happening in terms of resonance.

9. **Applying Concepts** How could two waves on a rope interfere so that the rope did not move at all?

Interpreting Graphics

10. In the image below, what sort of wave interaction is happening?

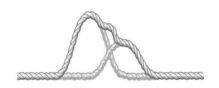

Skills Practice Lab

OBJECTIVES

Form hypotheses about the energy and speed of waves.

Test your hypotheses by performing an experiment.

MATERIALS

- beaker, small
- newspaper
- pan, shallow, approximately 20 cm × 30 cm
- pencils (2)
- stopwatch
- water

SAFETY

Wave Energy and Speed

If you threw a rock into a pond, waves would carry energy away from the point of origin. But if you threw a large rock into a pond, would the waves carry more energy away from the point of origin than waves caused by a small rock? And would a large rock make waves that move faster than waves made by a small rock? In this lab, you'll answer these questions.

Ask a Question

1 In this lab, you will answer the following questions: Do waves made by a large disturbance carry more energy than waves made by a small disturbance? Do waves created by a large disturbance travel faster than waves created by a small disturbance?

Form a Hypothesis

2 Write a few sentences that answer the questions above.

Test the Hypothesis

3 Place the pan on a few sheets of newspaper. Using the small beaker, fill the pan with water.

4 Make sure that the water is still. Tap the surface of the water with the eraser end of one pencil. This tap represents the small disturbance. Record your observations about the size of the waves that are made and the path they take.

5. Repeat step 4. This time, use the stopwatch to record the amount of time it takes for one of the waves to reach the side of the pan. Record your data.

6. Using two pencils at once, repeat steps 4 and 5. These taps represent the large disturbance. (Try to use the same amount of force to tap the water that you used with just one pencil.) Observe and record your results.

Analyze the Results

1. **Describing Events** Compare the appearance of the waves created by one pencil with that of the waves created by two pencils. Were there any differences in amplitude (wave height)?

2. **Describing Events** Compare the amount of time required for the waves to reach the side of the pan. Did the waves travel faster when two pencils were used?

Draw Conclusions

3. **Drawing Conclusions** Do waves made by a large disturbance carry more energy than waves made by a small one? Explain your answer, using your results to support your answer. (Hint: Remember the relationship between amplitude and energy.)

4. **Drawing Conclusions** Do waves made by a large disturbance travel faster than waves made by a small one? Explain your answer.

Applying Your Data

A tsunami is a giant ocean wave that can reach a height of 30 m. Tsunamis that reach land can cause injury and enormous property damage. Using what you learned in this lab about wave energy and speed, explain why tsunamis are so dangerous. How do you think scientists can predict when tsunamis will reach land?

Chapter Review

USING KEY TERMS

For each pair of terms, explain how the meanings of the terms differ.

1 *longitudinal wave* and *transverse wave*

2 *wavelength* and *amplitude*

3 *reflection* and *refraction*

UNDERSTANDING KEY IDEAS

Multiple Choice

4 As the wavelength increases, the frequency

  a. decreases.

  b. increases.

  c. remains the same.

  d. increases and then decreases.

5 Waves transfer

  a. matter.   **c.** particles.

  b. energy.   **d.** water.

6 Refraction occurs when a wave enters a new medium at an angle because

  a. the frequency changes.

  b. the amplitude changes.

  c. the wave speed changes.

  d. None of the above

7 The wave property that is related to the height of a wave is the

  a. wavelength.

  b. amplitude.

  c. frequency.

  d. wave speed.

8 During constructive interference,

  a. the amplitude increases.

  b. the frequency decreases.

  c. the wave speed increases.

  d. All of the above

9 Waves that don't require a medium are

  a. longitudinal waves.

  b. electromagnetic waves.

  c. surface waves.

  d. mechanical waves.

Short Answer

10 Draw a transverse wave and a longitudinal wave. Label a crest, a trough, a compression, a rarefaction, and wavelengths. Also, label the amplitude on the transverse wave.

11 What is the relationship between frequency, wave speed, and wavelength?

Math Skills

12 A fisherman in a row boat notices that one wave crest passes his fishing line every 5 s. He estimates the distance between the crests to be 1.5 m and estimates that the crests of the waves are 0.5 m above the troughs. Using this data, determine the amplitude and speed of the waves.

13 Concept Mapping Use the following terms to create a concept map: *wave, refraction, transverse wave, longitudinal wave, wavelength, wave speed,* and *diffraction.*

14 Analyzing Ideas You have lost the paddles for the canoe you rented, and the canoe has drifted to the center of a pond. You need to get it back to the shore, but you do not want to get wet by swimming in the pond. Your friend suggests that you drop rocks behind the canoe to create waves that will push the canoe toward the shore. Will this solution work? Why or why not?

15 Applying Concepts Some opera singers can use their powerful voices to break crystal glasses. To do this, they sing one note very loudly and hold it for a long time. While the opera singer holds the note, the walls of the glass move back and forth until the glass shatters. Explain in terms of resonance how the glass shatters.

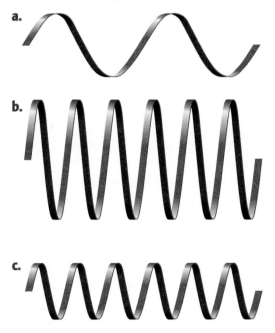

16 Analyzing Processes After setting up stereo speakers in your school's music room, you notice that in certain areas of the room, the sound from the speakers is very loud. In other areas, the sound is very soft. Using the concept of interference, explain why the sound levels in the music room vary.

17 Predicting Consequences A certain sound wave travels through water with a certain wavelength, frequency, and wave speed. A second sound wave with twice the frequency of the first wave then travels through the same water. What is the second wave's wavelength and wave speed compared to those of the first wave?

INTERPRETING GRAPHICS

18 Look at the waves below. Rank the waves from highest energy to lowest energy, and explain your reasoning.

a.

b.

c.

Standardized Test Preparation

Read each of the passages below. Then, answer the questions that follow each passage.

Passage 1 On March 27, 1964, a powerful earthquake rocked Alaska. The earthquake started on land near Anchorage, and the seismic waves spread quickly in all directions. The earthquake created a series of ocean waves called <u>tsunamis</u> in the Gulf of Alaska. In the deep water of the gulf, the tsunamis were short and far apart. But as these waves entered the shallow water surrounding Kodiak Island, off the coast of Alaska, they became taller and closer together. Some reached heights of nearly 30 m! The destructive forces of the earthquake and tsunamis killed 21 people and caused $10 million in damage to Kodiak, which made this marine disaster the worst in the town's 200-year history.

1. In the passage, what does *tsunami* mean?

A a seismic wave

B an earthquake

C an ocean wave

D a body of water

2. Which of these events happened first?

F The tsunamis became closer together.

G Tsunamis entered the shallow water.

H Tsunamis formed in the Gulf of Alaska.

I An earthquake began near Anchorage.

3. Which conclusion is **best** supported by information given in the passage?

A Kodiak had never experienced a tsunami before 1964.

B Tsunamis and an earthquake were the cause of Kodiak's worst marine disaster in 200 years.

C Tsunamis are common in Kodiak.

D The citizens of Kodiak went into debt after the 1964 earthquake.

Passage 2 Resonance was partially responsible for the destruction of the Tacoma Narrows Bridge, in Washington. The bridge opened in July 1940 and soon earned the nickname Galloping Gertie because of its wavelike motions. These motions were created by wind that blew across the bridge. The wind caused vibrations that were close to a resonant frequency of the bridge. Because the bridge was in resonance, it absorbed a large amount of energy from the wind, which caused it to vibrate with a large amplitude. On November 7, 1940, a supporting cable slipped, and the bridge began to twist. The twisting of the bridge, combined with high winds, further increased the amplitude of the bridge's motion. Within hours, the amplitude became so great that the bridge collapsed. Luckily, all of the people on the bridge that day were able to escape before it crashed into the river below.

1. What caused wavelike motions in the Tacoma Narrows Bridge?

A wind that caused vibrations that were close to the resonant frequency of the bridge

B vibrations from cars going over the bridge

C twisting of a broken support cable

D an earthquake

2. Why did the bridge collapse?

F A supporting cable slipped.

G It absorbed a great amount of energy from the wind.

H The amplitude of the bridge's vibrations became great enough.

I Wind blew across it.

Use the figure below to answer the questions that follow.

Read each question below, and choose the best answer.

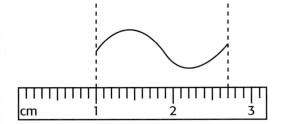

1. This wave was generated in a laboratory investigation. What is the wavelength of the wave?

A 1.5 cm

B 1.7 cm

C 2.0 cm

D 2.7 cm

2. If the frequency of the wave shown were doubled, what would the wavelength of the wave be?

F 0.85 cm

G 1.35 cm

H 3.4 cm

I 5.4 cm

3. What is the amplitude of the wave shown?

A 0.85 cm

B 1.7 cm

C 2.7 cm

D There is not enough information to determine the answer.

1. How is the product of $5 \times 5 \times 5 \times 2 \times 2 \times 2 \times 2$ expressed in exponential notation?

A $3^5 \times 4^2$

B $5^3 \times 2^4$

C $5^7 \times 2^7$

D 10^7

2. Mannie purchased 8.9 kg of dog food from the veterinarian. How many grams of dog food did he purchase?

F 8,900 g

G 890 g

H 89 g

I 0.89 g

3. What is the area of a rectangle whose sides are 3 cm long and 7.5 cm long?

A 10.5 cm²

B 12 cm²

C 21 cm²

D 22.5 cm²

4. An underwater sound wave traveled 1.5 km in 1 s. How far would it travel in 4 s?

F 5.0 km

G 5.5 km

H 6.0 km

I 6.5 km

5. During a tennis game, the person serving the ball is allowed only 2 serves to start a point. Hannah plays a tennis match and is able to use 50 of her 63 first serves to start a point. What is the **best** estimate of Hannah's first-service percentage?

A 126%

B 88%

C 81.5%

D 79%

Science in Action

Science, Technology, and Society

The Ultimate Telescope

The largest telescopes in the world don't depend on visible light, lenses, or mirrors. Instead, they collect radio waves from the far reaches of outer space. One radio telescope, called the Very Large Array (VLA), is located in a remote desert in New Mexico.

Just as you can detect light waves from stars with your eyes, radio waves emitted from objects in space can be detected with radio telescopes. The Very Large Array consists of 27 radio telescopes like the ones in the photo above.

Math ACTiViTY

Radio waves travel about 300,000,000 m/s. The M100 galaxy is about 5.68×10^{23} m away from Earth. How long, in years, does it take radio waves from M100 to be detected by the VLA?

Scientific Discoveries

The Wave Nature of Light

Have you ever wondered what light really is? Many early scientists did. One of them, the great 17th-century scientist Isaac Newton, did some experiments and decided that light consisted of particles. But when experimenting with lenses, Newton observed some things that he could not explain.

Around 1800, the scientist Thomas Young did more experiments on light and found that it diffracted when it passed through slits. Young concluded that light could be thought of as waves. Although scientists were slow to accept this idea, they now know that light is both particle-like and wavelike.

Language Arts ACTiViTY

WRITING SKILL Thomas Young said, "The nature of light is a subject of no material importance to the concerns of life or to the practice of the arts, but it is in many other respects extremely interesting." Write a brief essay in which you answer the following questions: What do you think Young meant? Do you agree with him? How would you respond to his statement?

Estela Zavala

Ultrasonographer Estela Zavala is a registered diagnostic medical ultrasonographer who works at Austin Radiological Association in Austin, Texas. Most people have seen a picture of a sonogram showing an unborn baby inside its mother's womb. Ultrasound technologists make these images with an ultrasound machine, which sends harmless, high-frequency sound waves into the body. Zavala uses ultrasound to form images of organs in the body. Zavala says about her education, "After graduating from high school, I went to an X-ray school to be licensed as an X-ray technologist. First, I went to an intensive one-month training program. After that, I worked for a licensed radiologist for about a year. Finally, I attended a year-long ultrasound program at a local community college before becoming fully licensed." What Zavala likes best about her job is being able to help people by finding out what is wrong with them without surgery. Before ultrasound, surgery was the only way to find out about the health of someone's organs.

Social Studies ACTiViTY

WRITING SKILL Research the different ways in which ultrasound technology is used in medical practice today. Write a few paragraphs about what you learn.

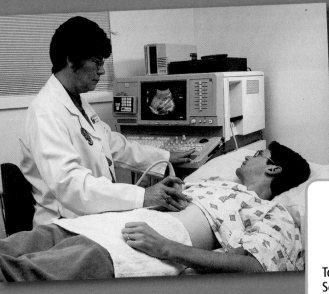

go.hrw.com

To learn more about these Science in Action topics, visit **go.hrw.com** and type in the keyword **HP5WAVF**.

Current Science

Check out Current Science® articles related to this chapter by visiting go.hrw.com. Just type in the keyword **HP5CS20**.

2

The Nature of Sound

The Big Idea

The properties and interactions of sound waves affect what one hears.

About the PHOTO

Look at these dolphins swimming swiftly and silently through their watery world. Wait a minute—swiftly? Yes. Silently? No way! Dolphins use sound—clicks, squeaks, and other noises—to communicate. Dolphins also use sound to locate their food by echolocation and to find their way through murky water.

PRE-READING ACTIVITY

Graphic Organizer

Concept Map Before you read the chapter, create the graphic organizer entitled "Concept Map" described in the **Study Skills** section of the Appendix. As you read the chapter, fill in the concept map with details about each type of sound interaction.

START-UP ACTIVITY

A Homemade Guitar

In this chapter, you will learn about sound. You can start by making your own guitar. It won't sound as good as a real guitar, but it will help you explore the nature of sound.

Procedure

1. Stretch a **rubber band** lengthwise around an empty **shoe box.** Place the box hollow side up. Pluck the rubber band gently. Describe what you hear.

2. Stretch **another rubber band of a different thickness** around the box. Pluck both rubber bands. Describe the differences in the sounds.

3. Put a **pencil** across the center of the box and under the rubber bands, and pluck again. Compare this sound with the sound you heard before the pencil was used.

4. Move the pencil closer to one end of the shoe box. Pluck on both sides of the pencil. Describe the differences in the sounds you hear.

Analysis

1. How did the thicknesses of the rubber bands affect the sound?

2. In steps 3 and 4, you changed the length of the vibrating part of the rubber bands. What is the relationship between the vibrating length of the rubber band and the sound that you hear?

What Is Sound?

You are in a restaurant, and without warning, you hear a loud crash. A waiter dropped a tray of dishes. What a mess! But why did dropping the dishes make such a loud sound?

In this section, you'll find out what causes sound and what characteristics all sounds have in common. You'll also learn how your ears detect sound and how you can protect your hearing.

Sound and Vibrations

As different as they are, all sounds have some things in common. One characteristic of sound is that it is created by vibrations. A *vibration* is the complete back-and-forth motion of an object. **Figure 1** shows one way sound is made by vibrations.

What You Will Learn

- Describe how vibrations cause sound.
- Explain how sound is transmitted through a medium.
- Explain how the human ear works, and identify its parts.
- Identify ways to protect your hearing.

Vocabulary

sound wave
medium

READING STRATEGY

Prediction Guide Before reading this section, predict whether each of the following statements is true or false:

- Sound waves are made by vibrations.
- Sound waves push air particles along until they reach your ear.

Figure 1 Sounds from a Stereo Speaker

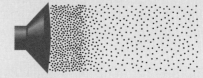

ⓐ Electrical signals make the speaker vibrate. As the speaker cone moves forward, it pushes the air particles in front of it closer together, creating a region of higher density and pressure called a *compression*.

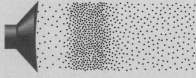

ⓑ As the speaker cone moves backward, air particles close to the cone become less crowded, creating a region of lower density and pressure called a *rarefaction*.

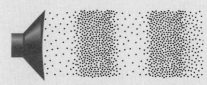

ⓒ For each vibration, a compression and a rarefaction are formed. As the compressions and rarefactions travel away from the speaker, sound is transmitted through the air.

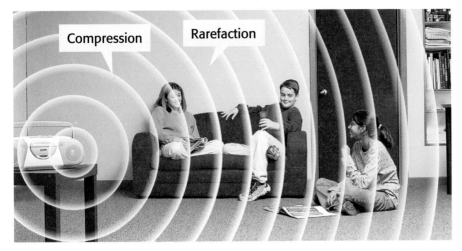

Compression   Rarefaction

Figure 2 *You can't actually see sound waves, but they can be represented by spheres that spread out in all directions.*

Sound Waves

Longitudinal (LAHN juh TOOD'n uhl) waves are made of compressions and rarefactions. A **sound wave** is a longitudinal wave caused by vibrations and carried through a substance. The particles of the substance, such as air particles, vibrate back and forth along the path that the sound wave travels. Sound is transmitted through the vibrations and collisions of the particles. Because the particles vibrate back and forth along the paths that sound travels, sound travels as longitudinal waves.

Sound waves travel in all directions away from their source, as shown in **Figure 2.** However, air or other matter does not travel with the sound waves. The particles of air only vibrate back and forth. If air did travel with sound, wind gusts from music speakers would blow you over at a school dance!

sound wave a longitudinal wave that is caused by vibrations and that travels through a material medium

☑ **Reading Check** What do sound waves consist of? (*See the Appendix for answers to Reading Checks.*)

Good Vibrations

1. Gently strike a **tuning fork** on a **rubber eraser.** Watch the prongs, and listen for a sound. Describe what you see and what you hear.
2. Lightly touch the fork with your fingers. What do you feel?
3. Grasp the prongs of the fork firmly with your hand. What happens to the sound?
4. Strike the tuning fork on the eraser again, and dip the prongs in a **cup of water.** Describe what happens to the water.
5. Record your observations.

Figure 3 Tubing is connected to a pump that is removing air from the jar. As the air is removed, the ringing alarm clock sounds quieter and quieter.

medium a physical environment in which phenomena occur

CONNECTION TO Biology

Vocal Sounds The vibrations that produce your voice are made inside your throat. When you speak, laugh, or sing, your lungs force air up your windpipe, causing your vocal cords to vibrate.

Do some research, and find out what role different parts of your throat and mouth play in making vocal sounds. Make a poster in which you show the different parts, and explain the role they play in shaping sound waves.

ACTIVITY

Sound and Media

Another characteristic of sound is that all sound waves require a medium (plural, *media*). A **medium** is a substance through which a wave can travel. Most of the sounds that you hear travel through air at least part of the time. But sound waves can also travel through other materials, such as water, glass, and metal.

In a vacuum, however, there are no particles to vibrate. So, no sound can be made in a vacuum. This fact helps to explain the effect described in **Figure 3.** Sound must travel through air or some other medium to reach your ears and be detected.

✓ **Reading Check** What does sound need in order to travel?

How You Detect Sound

Imagine that you are watching a suspenseful movie. Just before a door is opened, the background music becomes louder. You know that there is something scary behind that door! Now, imagine watching the same scene without the sound. You would have more difficulty figuring out what's going on if there were no sound.

Figure 4 shows how your ears change sound waves into electrical signals that allow you to hear. First, the outer ear collects sound waves. The vibrations then go to your middle ear. Very small organs increase the size of the vibrations here. These vibrations are then picked up by organs in your inner ear. Your inner ear changes vibrations into electrical signals that your brain interprets as sound.

Figure 4 **How the Human Ear Works**

ⓐ The **outer ear** acts as a funnel for sound waves. The *pinna* collects sound waves and directs them into the *ear canal.*

ⓑ In the **middle ear,** three bones—the *hammer, anvil,* and *stirrup*—act as levers to increase the size of the vibrations.

ⓒ In the **inner ear,** vibrations created by sound are changed into electrical signals for the brain to interpret.

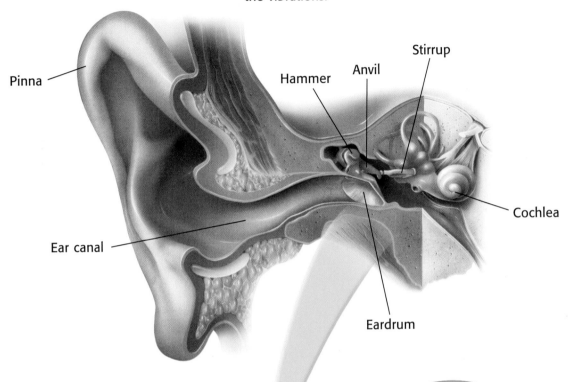

Stirrup

Anvil

Hammer

Pinna

Cochlea

Ear canal

Eardrum

❶ Sound waves vibrate the *eardrum*—a lightly stretched membrane that is the entrance to the middle ear.

❷ The vibration of the eardrum makes the hammer vibrate, which, in turn, makes the anvil and stirrup vibrate.

❸ The stirrup vibrates the *oval window*—the entrance to the inner ear.

❹ The vibrations of the oval window create waves in the liquid inside the *cochlea.*

❺ Movement of the liquid causes tiny hair cells inside the cochlea to bend.

❻ The bending of the hair cells stimulates nerves, which send electrical signals to the brain.

Figure 5 *Sound is made whether or not anyone is around to hear it.*

INTERNET ACTIVITY

For another activity related to this chapter, go to **go.hrw.com** and type in the keyword **HP5SNDW.**

Making Sound Versus Hearing Sound

Have you heard this riddle? If a tree falls in the forest and no one is around to hear it, does the tree make a sound? Think about the situation pictured in **Figure 5.** When a tree falls and hits the ground, the tree and the ground vibrate. These vibrations make compressions and rarefactions in the surrounding air. So, there would be a sound!

Making sound is separate from detecting sound. The fact that no one heard the tree fall doesn't mean that there wasn't a sound. A sound was made—it just wasn't heard.

Hearing Loss and Deafness

The many parts of the ear must work together for you to hear sounds. If any part of the ear is damaged or does not work properly, hearing loss or deafness may result.

One of the most common types of hearing loss is called *tinnitus* (ti NIET us), which results from long-term exposure to loud sounds. Loud sounds can cause damage to the hair cells and nerve endings in the cochlea. Once these hairs are damaged, they do not grow back. Damage to the cochlea or any other part of the inner ear usually results in permanent hearing loss.

People who have tinnitus often say they have a ringing in their ears. They also have trouble understanding other people and hearing the difference between words that sound alike. Tinnitus can affect people of any age. Fortunately, tinnitus can be prevented.

Reading Check What causes tinnitus?

Protecting Your Hearing

Short exposures to sounds that are loud enough to be painful can cause hearing loss. Your hearing can also be damaged by loud sounds that are not quite painful, if you are exposed to them for long periods of time. There are some simple things you can do to protect your hearing. Loud sounds can be blocked out by earplugs. You can listen at a lower volume when you are using headphones, as in **Figure 6.** You can also move away from loud sounds. If you are near a speaker playing loud music, just move away from it. When you double the distance between yourself and a loud sound, the sound's intensity to your ears will be one-fourth of what it was before.

Figure 6 *Turning your radio down can help prevent hearing loss, especially when you use headphones.*

SECTION Review

Summary

- All sounds are generated by vibrations.
- Sounds travel as longitudinal waves consisting of compressions and rarefactions.
- Sound waves travel in all directions away from their source.
- Sound waves require a medium through which to travel. Sound cannot travel in a vacuum.
- Your ears convert sound into electrical impulses that are sent to your brain.
- Exposure to loud sounds can cause hearing damage.
- Using earplugs and lowering the volume of sounds can prevent hearing damage.

Using Key Terms

1. Use the following terms in the same sentence: *sound wave* and *medium*.

Understanding Key Ideas

2. Sound travels as
   a. transverse waves.
   b. longitudinal waves.
   c. shock waves.
   d. airwaves.

3. Which part of the ear increases the size of the vibrations of sound waves entering the ear?
   a. outer ear
   b. ear canal
   c. middle ear
   d. inner ear

4. Name two ways of protecting your hearing.

Critical Thinking

5. **Analyzing Processes** Explain why a person at a rock concert will not feel gusts of wind coming out of the speakers.

6. **Analyzing Ideas** If a meteorite crashed on the moon, would you be able to hear it on Earth? Why, or why not?

7. **Identifying Relationships** Recall the breaking dishes mentioned at the beginning of this section. Why was the sound that they made so loud?

Interpreting Graphics

Use the diagram of a wave below to answer the questions that follow.

8. What kind of wave is this?

9. Draw a sketch of the diagram on a separate sheet of paper, and label the compressions and rarefactions.

10. How do vibrations make these kinds of waves?

Properties of Sound

Imagine that you are swimming in a neighborhood pool. You can hear the high, loud laughter of small children and the soft splashing of the waves at the edge of the pool.

Why are some sounds loud, soft, high, or low? The differences between sounds depend on the properties of the sound waves. In this section, you will learn about properties of sound.

The Speed of Sound

Suppose you are standing at one end of a pool and two people from the opposite end of the pool yell at the same time. You would hear their voices at the same time. The reason is that the speed of sound depends only on the medium in which the sound is traveling. So, you would hear them at the same time—even if one person yelled louder!

How the Speed of Sound Can Change

Table 1 shows how the speed of sound varies in different media. Sound travels quickly through air, but it travels even faster in liquids and even faster in solids.

Temperature also affects the speed of sound. In general, the cooler the medium is, the slower the speed of sound. Particles of cool materials move more slowly and transmit energy more slowly than particles do in warmer materials. In 1947, pilot Chuck Yeager became the first person to travel faster than the speed of sound. Yeager flew the airplane shown in **Figure 1** at 293 m/s (about 480 mi/h) at 12,000 m above sea level. At that altitude, the temperature of the air is so low that the speed of sound is only 290 m/s.

Table 1 Speed of Sound in Different Media

Medium	Speed (m/s)
Air (0°C)	331
Air (20°C)	343
Air (100°C)	366
Water (20°C)	1,482
Steel (20°C)	5,200

Figure 1 The X-1 airplane was the first vehicle to move faster than the speed of sound.

Figure 2 Frequency and Pitch

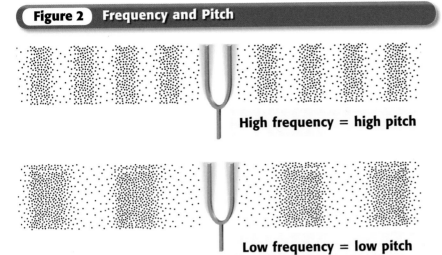

High frequency = high pitch

Low frequency = low pitch

Pitch and Frequency

How low or high a sound seems to be is the **pitch** of that sound. The *frequency* of a wave is the number of crests or troughs that are made in a given time. The pitch of a sound is related to the frequency of the sound wave, as shown in **Figure 2.** Frequency is expressed in hertz (Hz), where 1 Hz = 1 wave per second. For example, the lowest note on a piano is about 40 Hz. The screech of a bat is 10,000 Hz or higher.

✓ Reading Check What is frequency? (*See the Appendix for answers to Reading Checks.*)

Frequency and Hearing

If you see someone blow a dog whistle, the whistle seems silent to you. The reason is that the frequency of the sound wave is out of the range of human hearing. But the dog hears the whistle and comes running! **Table 2** compares the range of frequencies that humans and animals can hear. Sounds that have a frequency too high for people to hear are called *ultrasonic.*

pitch a measure of how high or low a sound is perceived to be, depending on the frequency of the sound wave

The Speed of Sound

The speed of sound depends on the medium through which sound is traveling and the medium's temperature. Sound travels at 343 m/s through air at a temperature of 20°C. How far will sound travel in 20°C air in 5 s?

The speed of sound in steel at 20°C is 5,200 m/s. How far can sound travel in 5 s through steel at 20°C?

Table 2 Frequencies Heard by Different Animals	
Animal	**Frequency range (Hz)**
Bat	2,000 to 110,000
Porpoise	75 to 150,000
Cat	45 to 64,000
Beluga whale	1,000 to 123,000
Elephant	16 to 12,000
Human	20 to 20,000
Dog	67 to 45,000

Figure 3 The Doppler Effect

a A car with its horn honking moves toward the sound waves going in the same direction. A person in front of the car hears sound waves that are closer together.

b The car moves away from the sound waves going in the opposite direction. A person behind the car hears sound waves that are farther apart and have a lower frequency.

The Doppler Effect

Have you ever been passed by a car with its horn honking? If so, you probably noticed the sudden change in pitch—sort of an *EEEEEOOooooowwn* sound—as the car went past you. The pitch you heard was higher as the car moved toward you than it was after the car passed. This higher pitch was a result of the Doppler effect. For sound waves, the **Doppler effect** is the apparent change in the frequency of a sound caused by the motion of either the listener or the source of the sound. **Figure 3** shows how the Doppler effect works.

In a moving sound source, such as a car with its horn honking, sound waves that are moving forward are going the same direction the car is moving. As a result, the compressions and rarefactions of the sound wave will be closer together than they would be if the sound source was not moving. To a person in front of the car, the frequency and pitch of the sound seem high. After the car passes, it is moving in the opposite direction that the sound waves are moving. To a person behind the car, the frequency and pitch of the sound seem low. The driver always hears the same pitch because the driver is moving with the car.

Doppler effect an observed change in the frequency of a wave when the source or observer is moving

Loudness and Amplitude

If you gently tap a drum, you will hear a soft rumbling. But if you strike the drum with a large force, you will hear a much louder sound! By changing the force you use to strike the drum, you change the loudness of the sound that is created. **Loudness** is a measure of how well a sound can be heard.

Energy and Vibration

Look at **Figure 4.** The harder you strike a drum, the louder the boom. As you strike the drum harder, you transfer more energy to the drum. The drum moves with a larger vibration and transfers more energy to the air around it. This increase in energy causes air particles to vibrate farther from their rest positions.

Increasing Amplitude

When you strike a drum harder, you are increasing the amplitude of the sound waves being made. The *amplitude* of a wave is the largest distance the particles in a wave vibrate from their rest positions. The larger the amplitude, the louder the sound. And the smaller the amplitude, the softer the sound. One way to increase the loudness of a sound is to use an amplifier, shown in **Figure 5.** An amplifier receives sound signals in the form of electric current. The amplifier then increases the energy and makes the sound louder.

✔ Reading Check What is the relationship between the amplitude of a sound and its energy of vibration?

Figure 4 *When a drum is struck hard, it vibrates with a lot of energy, making a loud sound.*

loudness the extent to which a sound can be heard

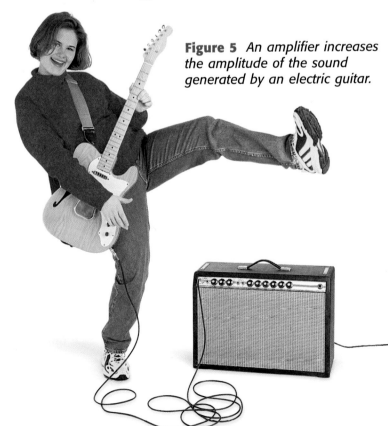

Figure 5 *An amplifier increases the amplitude of the sound generated by an electric guitar.*

Quick Lab

Sounding Board

1. With one hand, hold a **ruler** on your **desk** so that one end of it hangs over the edge.

2. With your other hand, pull the free end of the ruler up a few centimeters, and let go.

3. Try pulling the ruler up different distances. How does the distance affect the sounds you hear? What property of the sound wave are you changing?

4. Change the length of the part that hangs over the edge. What property of the sound wave is affected? Record your answers and observations.

Table 3 Decibel Levels of Common Sounds

Decibel level	Sound
0	the softest sounds you can hear
20	whisper
25	purring cat
60	normal conversation
80	lawn mower, vacuum cleaner, truck traffic
100	chain saw, snowmobile
115	sandblaster, loud rock concert, automobile horn
120	threshold of pain
140	jet engine 30 m away
200	rocket engine 50 m away

decibel the most common unit used to measure loudness (symbol, dB)

Measuring Loudness

The most common unit used to express loudness is the **decibel** (dB). The softest sounds an average human can hear are at a level of 0 dB. Sounds that are at 120 dB or higher can be painful. **Table 3** shows some common sounds and their decibel levels.

"Seeing" Amplitude and Frequency

Sound waves are invisible. However, technology can provide a way to "see" sound waves. A device called an *oscilloscope* (uh SIL uh SKOHP) can graph representations of sound waves, as shown in **Figure 6.** Notice that the graphs look like transverse waves instead of longitudinal waves.

Reading Check What does an oscilloscope do?

Figure 6 "Seeing" Sounds

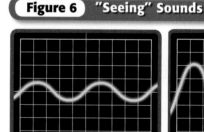
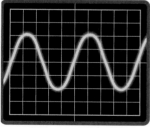
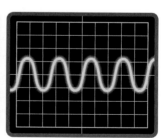

The graph on the right has a **larger amplitude** than the graph on the left. So, the sound represented on the right is **louder** than the one represented on the left.

The graph on the right has a **lower frequency** than the one on the left. So, the sound represented on the right has a **lower pitch** than the one represented on the left.

From Sound to Electrical Signal

An oscilloscope is shown in **Figure 7.** A microphone is attached to the oscilloscope and changes a sound wave into an electrical signal. The electrical signal is graphed on the screen in the form of a wave. The graph shows the sound as if it were a transverse wave. So, the sound's amplitude and frequency are easier to see. The highest points (crests) of these waves represent compressions, and the lowest points (troughs) represent rarefactions. By looking at the displays on the oscilloscope, you can quickly see the differences in amplitude and frequency of different sound waves.

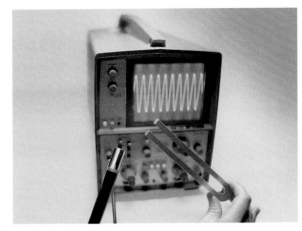

Figure 7 *An oscilloscope can be used to represent sounds.*

SECTION Review

Summary

- The speed of sound depends on the medium and the temperature.
- The pitch of a sound becomes higher as the frequency of the sound wave becomes higher. Frequency is expressed in units of Hertz (Hz), which is equivalent to waves per second.
- The Doppler effect is the apparent change in frequency of a sound caused by the motion of either the listener or the source of the sound.
- Loudness increases with the amplitude of the sound. Loudness is expressed in decibels.
- The amplitude and frequency of a sound can be measured electronically by an oscilloscope.

Using Key Terms

1. In your own words, write a definition for the term *pitch*.

2. Use the following terms in the same sentence: *loudness* and *decibel*.

Understanding Key Ideas

3. At the same temperature, in which medium does sound travel fastest?
   a. air
   b. liquid
   c. solid
   d. It travels at the same speed through all media.

4. In general, how does the temperature of a medium affect the speed of sound through that medium?

5. What property of waves affects the pitch of a sound?

6. How does an oscilloscope allow sound waves to be "seen"?

Math Skills

7. You see a distant flash of lightning, and then you hear a thunderclap 2 s later. The sound of the thunder moves at 343 m/s. How far away was the lightning?

8. In water that is near 0°C, a submarine sends out a sonar signal (a sound wave). The signal travels 1500 m/s and reaches an underwater mountain in 4 s. How far away is the mountain?

Critical Thinking

9. **Analyzing Processes** Will a listener notice the Doppler effect if both the listener and the source of the sound are traveling toward each other? Explain your answer.

10. **Predicting Consequences** A drum is struck gently, then is struck harder. What will be the difference in the amplitude of the sounds made? What will be the difference in the frequency of the sounds made?

Interactions of Sound Waves

Have you ever heard of a sea canary? It's not a bird! It's a whale! Beluga whales are sometimes called sea canaries because of the many different sounds they make.

Dolphins, beluga whales, and many other animals that live in the sea use sound to communicate. Beluga whales also rely on reflected sound waves to find fish, crabs, and shrimp to eat. In this section, you will learn about reflection and other interactions of sound waves. You will also learn how bats, dolphins, and whales use sound to find food.

Reflection of Sound Waves

Reflection is the bouncing back of a wave after it strikes a barrier. You're probably already familiar with a reflected sound wave, otherwise known as an **echo.** The strength of a reflected sound wave depends on the reflecting surface. Sound waves reflect best off smooth, hard surfaces. Look at **Figure 1.** A shout in an empty gymnasium can produce an echo, but a shout in an auditorium usually does not.

The difference is that the walls of an auditorium are usually designed so that they absorb sound. If sound waves hit a flat, hard surface, they will reflect back. Reflection of sound waves doesn't matter much in a gymnasium. But you don't want to hear echoes while listening to a musical performance!

What You Will Learn

- Explain how echoes are made, and describe their use in locating objects.
- List examples of constructive and destructive interference of sound waves.
- Explain what resonance is.

Vocabulary

echo
echolocation
interference
sonic boom
standing wave
resonance

echo a reflected sound wave

Figure 1   Sound Reflection and Absorption

Sound waves easily reflect off the smooth, hard walls of a gymnasium. For this reason, you hear an echo.

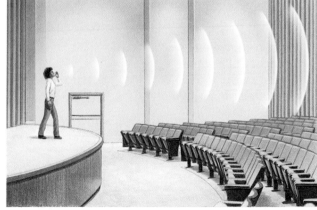

In well-designed auditoriums, echoes are reduced by soft materials that absorb sound waves and by irregular shapes that scatter sound waves.

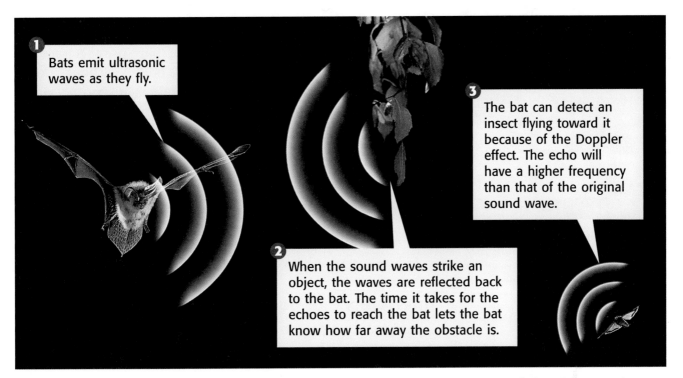

Figure 2 *Bats use echolocation to navigate around barriers and to find insects to eat.*

Echolocation

Beluga whales use echoes to find food. The use of reflected sound waves to find objects is called **echolocation.** Other animals—such as dolphins, bats, and some kinds of birds—also use echolocation to hunt food and to find objects in their paths. **Figure 2** shows how echolocation works. Animals that use echolocation can tell how far away something is based on how long it takes sound waves to echo back to their ears. Some animals, such as bats, also make use of the Doppler effect to tell if another moving object, such as an insect, is moving toward it or away from it.

✓ **Reading Check** How is echolocation useful to some animals? (*See the Appendix for answers to Reading Checks.*)

Echolocation Technology

People use echoes to locate objects underwater by using sonar (which stands for **s**ound **n**avigation **a**nd **r**anging). *Sonar* is a type of electronic echolocation. **Figure 3** shows how sonar works. Ultrasonic waves are used because their short wavelengths give more details about the objects they reflect off. Sonar can also help navigators on ships avoid icebergs and can help oceanographers map the ocean floor.

echolocation the process of using reflected sound waves to find objects; used by animals such as bats

Figure 3 *A fish finder sends ultrasonic waves down into the water. The time it takes for the echo to return helps determine the location of the fish.*

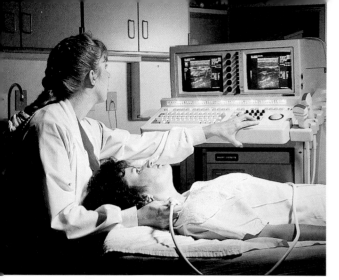

Figure 4 *Images created by ultrasonography are fuzzy, but they are a safe way to see inside a patient's body.*

Ultrasonography

Ultrasonography (UHL truh soh NAHG ruh fee) is a medical procedure that uses echoes to "see" inside a patient's body without doing surgery. A special device makes ultrasonic waves with a frequency that can be from 1 million to 10 million hertz, which reflect off the patient's internal organs. These echoes are then changed into images that can be seen on a television screen, as shown in **Figure 4.** Ultrasonography is used to examine kidneys, gallbladders, and other organs. It is also used to check the development of an unborn baby in a mother's body. Ultrasonic waves are less harmful to human tissue than X rays are.

Interference of Sound Waves

interference the combination of two or more waves that results in a single wave

sonic boom the explosive sound heard when a shock wave from an object traveling faster than the speed of sound reaches a person's ears

Sound waves also interact through interference. **Interference** happens when two or more waves overlap. **Figure 5** shows how two sound waves can combine by both constructive and destructive interference.

Orchestras and bands make use of constructive interference when several instruments of the same kind play the same notes. Interference of the sound waves causes the combined amplitude to increase, resulting in a louder sound. But destructive interference may keep some members of the audience from hearing the concert well. In certain places in an auditorium, sound waves reflecting off the walls interfere destructively with the sound waves from the stage.

✓ Reading Check What are the two kinds of sound wave interference?

Figure 5 **Constructive and Destructive Interference**

Sound waves from two speakers producing sound of the same frequency combine by both constructive and destructive interference.

Constructive Interference
As the compressions of one wave overlap the compressions of another wave, the sound will be louder because the amplitude is increased.

Destructive Interference
As the compressions of one wave overlap the rarefactions of another wave, the sound will be softer because the amplitude is decreased.

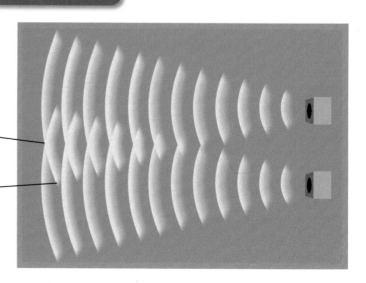

Interference and the Sound Barrier

As the source of a sound—such as a jet plane—gets close to the speed of sound, the sound waves in front of the jet plane get closer and closer together. The result is constructive interference. **Figure 6** shows what happens as a jet plane reaches the speed of sound.

For the jet in **Figure 6** to go faster than the speed of sound, the jet must overcome the pressure of the compressed sound waves. **Figure 7** shows what happens as soon as the jet reaches supersonic speeds—speeds faster than the speed of sound. At these speeds, the sound waves trail off behind the jet. At their outer edges, the sound waves combine by constructive interference to form a *shock wave*.

A **sonic boom** is the explosive sound heard when a shock wave reaches your ears. Sonic booms can be so loud that they can hurt your ears and break windows. They can even make the ground shake as it does during an earthquake.

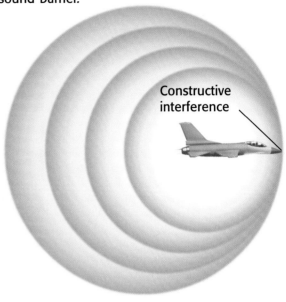

Figure 6 *When a jet plane reaches the speed of sound, the sound waves in front of the jet combine by constructive interference. The result is a high-density compression that is called the sound barrier.*

Constructive interference

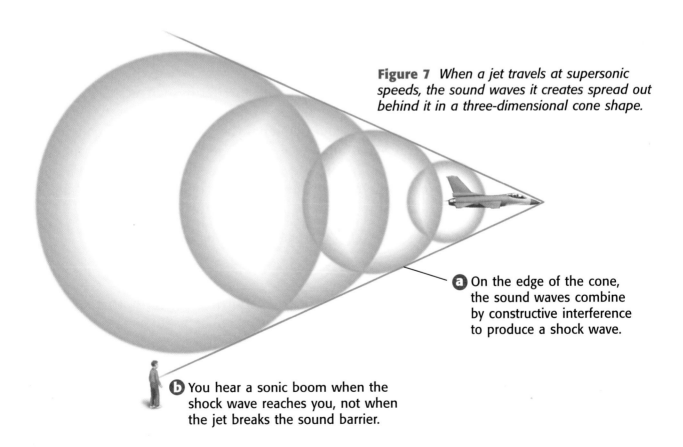

Figure 7 *When a jet travels at supersonic speeds, the sound waves it creates spread out behind it in a three-dimensional cone shape.*

a On the edge of the cone, the sound waves combine by constructive interference to produce a shock wave.

b You hear a sonic boom when the shock wave reaches you, not when the jet breaks the sound barrier.

Figure 8   **Resonant Frequencies of a Plucked String**

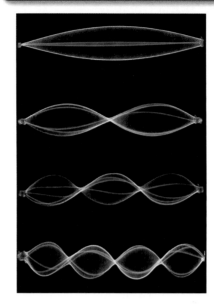

The lowest resonant frequency is called the *fundamental.*

Higher resonant frequencies are called *overtones.* The first overtone is twice the frequency of the fundamental.

The second overtone is 3 times the fundamental.

The third overtone is 4 times the fundamental.

standing wave a pattern of vibration that simulates a wave that is standing still

resonance a phenomenon that occurs when two objects naturally vibrate at the same frequency; the sound produced by one object causes the other object to vibrate

Interference and Standing Waves

When you play a guitar, you can make some pleasing sounds, and you might even play a tune. But have you ever watched a guitar string after you've plucked it? You may have noticed that the string vibrates as a standing wave. A **standing wave** is a pattern of vibration that looks like a wave that is standing still. Waves and reflected waves of the same frequency are going through the string. Where you see maximum amplitude, waves are interfering constructively. Where the string seems to be standing still, waves are interfering destructively.

Although you can see only one standing wave, which is at the *fundamental* frequency, the guitar string actually creates several standing waves of different frequencies at the same time. The frequencies at which standing waves are made are called *resonant frequencies.* Resonant frequencies and the relationships between them are shown in **Figure 8.**

✓ **Reading Check**   What is a standing wave?

Resonance

If you have a tuning fork, shown in **Figure 9,** that vibrates at one of the resonant frequencies of a guitar string, you can make the string make a sound without touching it. Strike the tuning fork, and hold it close to the string. The string will start to vibrate and produce a sound.

Using the vibrations of the tuning fork to make the string vibrate is an example of resonance. **Resonance** happens when an object vibrating at or near a resonant frequency of a second object causes the second object to vibrate.

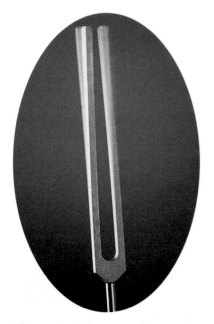

Figure 9 *When struck, a tuning fork can make another object vibrate if they both have the same resonant frequency.*

Resonance in Musical Instruments

Musical instruments use resonance to make sound. In wind instruments, vibrations are caused by blowing air into the mouthpiece. The vibrations make a sound, which is amplified when it forms a standing wave inside the instrument.

String instruments also resonate when they are played. An acoustic guitar, such as the one shown in **Figure 10,** has a hollow body. When the strings vibrate, sound waves enter the body of the guitar. Standing waves form inside the body of the guitar, and the sound is amplified.

Figure 10 *The body of a guitar resonates when the guitar is strummed.*

SECTION
Review

Summary

- Echoes are reflected sound waves.
- Some animals can use echolocation to find food or to navigate around objects.
- People use echolocation technology in many underwater applications.
- Ultrasonography uses sound reflection for medical applications.
- Sound barriers and shock waves are created by interference.
- Standing waves form at an object's resonant frequencies.
- Resonance happens when a vibrating object causes a second object to vibrate at one of its resonant frequencies.

Using Key Terms

1. Use the following terms in the same sentence: *echo* and *echolocation.*

Complete each of the following sentences by choosing the correct term from the word bank.

> interference   standing wave
> sonic boom   resonance

2. When you pluck a string on a musical instrument, a(n) _____ forms.

3. When a vibrating object causes a nearby object to vibrate, _____ results.

Understanding Key Ideas

4. What causes an echo?
   a. reflection
   b. resonance
   c. constructive interference
   d. destructive interference

5. Describe a place in which you would expect to hear echoes.

6. How do bats use echoes to find insects to eat?

7. Give one example each of constructive and destructive interference of sound waves.

Math Skills

8. Sound travels through air at 343 m/s at 20°C. A bat emits an ultrasonic squeak and hears the echo 0.05 s later. How far away was the object that reflected it? (Hint: Remember that the sound must travel *to* the object and *back to* the bat.)

Critical Thinking

9. **Applying Concepts** Your friend is playing a song on a piano. Whenever your friend hits a certain key, the lamp on top of the piano rattles. Explain why the lamp rattles.

10. **Making Comparisons** Compare sonar and ultrasonography in locating objects.

SCiLINKS®

NSTA
Developed and maintained by the National Science Teachers Association

For a variety of links related to this chapter, go to www.scilinks.org

Topic: Interactions of Sound Waves
SciLinks code: HSM0804

47

Sound Quality

Have you ever been told that the music you really like is just a lot of noise? If you have, you know that people can disagree about the difference between noise and music.

You might think of noise as sounds you don't like and music as sounds that are pleasant to hear. But the difference between music and noise does not depend on whether you like the sound. The difference has to do with sound quality.

What You Will Learn

● Explain why different instruments have different sound qualities.
● Describe how each family of musical instruments produces sound.
● Explain how noise is different from music.

Vocabulary

sound quality
noise

READING STRATEGY

Reading Organizer As you read this section, make a table comparing the way different instruments produce sound.

What Is Sound Quality?

Imagine that the same note is played on a piano and on a violin. Could you tell the instruments apart without looking? The notes played have the same frequency. But you could probably tell them apart because the instruments make different sounds. The notes sound different because a single note on an instrument actually comes from several different pitches: the fundamental and several overtones. The result of the combination of these pitches is shown in **Figure 1.** The result of several pitches mixing together through interference is **sound quality.** Each instrument has a unique sound quality. **Figure 1** also shows how the sound quality differs when two instruments play the same note.

Figure 1 *Each instrument has a unique sound quality that results from the particular blend of overtones that it has.*

Fundamental

First overtone

Second overtone

Resulting sound

Piano

Violin

Sound Quality of Instruments

The difference in sound quality among different instruments comes from their structural differences. All instruments produce sound by vibrating. But instruments vary in the part that vibrates and in the way that the vibrations are made. There are three main families of instruments: string instruments, wind instruments, and percussion instruments.

sound quality the result of the blending of several pitches through interference

✓ Reading Check How do musical instruments differ in how they produce sound? (*See the Appendix for answers to Reading Checks.*)

String Instruments

Violins, guitars, and banjos are examples of string instruments. They make sound when their strings vibrate after being plucked or bowed. **Figure 2** shows how two different string instruments produce sounds.

Figure 2 String Instruments

ⓐ Cellos and guitars have strings of different thicknesses. The thicker the string is, the lower the pitch is.

ⓑ The pitch of the string can be changed by pushing the string against the neck of the instrument to change the string's length. Shorter strings vibrate at higher frequencies.

ⓒ A string vibrates when a bow is pulled across it or when the string is plucked.

ⓕ Pickups on the guitar convert the vibration of the guitar string into an electrical signal.

ⓓ The vibrations in the cello string make the bridge vibrate, which, in turn, makes the body of the cello vibrate.

ⓖ An amplifier converts the electrical signal back into a sound wave and increases the loudness of the sound.

ⓔ The body of the cello and the air inside it resonate with the string's vibration, creating a louder sound.

Figure 3   Wind Instruments

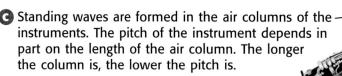

a A trumpet player's lips vibrate when the player blows into a trumpet.

b The reed vibrates back and forth when a musician blows into a clarinet.

c Standing waves are formed in the air columns of the instruments. The pitch of the instrument depends in part on the length of the air column. The longer the column is, the lower the pitch is.

d The length of the air column in a trumpet is changed by pushing the valves.

e The length of the air column in a clarinet is changed by closing or opening the finger holes.

Wind Instruments

A wind instrument produces sound when a vibration is created at one end of its air column. The vibration causes standing waves inside the air column. Pitch is changed by changing the length of the air column. Wind instruments are sometimes divided into two groups—woodwinds and brass. Examples of woodwinds are saxophones, oboes, and recorders. French horns, trombones, and tubas are brass instruments. A brass instrument and a woodwind instrument are shown in **Figure 3.**

Percussion Instruments

Drums, bells, and cymbals are percussion instruments. They make sound when struck. Instruments of different sizes are used to get different pitches. Usually, the larger the instrument is, the lower the pitch is. The drums and cymbals in a trap set, shown in **Figure 4,** are percussion instruments.

Figure 4   Percussion Instruments

The skins of the drums vibrate when struck with drumsticks.

Cymbals vibrate when struck together or when struck with drumsticks.

Each drum in the set is a different size. The larger the drum is, the lower the pitch is.

Music or Noise?

Most of the sounds we hear are noises. The sound of a truck roaring down the highway, the slam of a door, and the jingle of keys falling to the floor are all noises. **Noise** can be described as any sound, especially a nonmusical sound, that is a random mix of frequencies (or pitches). **Figure 5** shows on an oscilloscope the difference between a musical sound and noise.

noise a sound that consists of a random mix of frequencies

Reading Check What is the difference between music and noise?

French horn

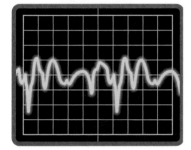

A sharp clap

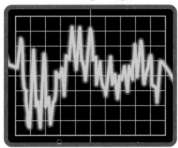

Figure 5 *A note from a French horn produces a sound wave with a repeating pattern, but noise from a clap produces complex sound waves with no regular pattern.*

SECTION Review

Summary

- Different instruments have different sound qualities.
- Sound quality results from the blending through interference of the fundamental and several overtones.
- The three families of instruments are string, wind, and percussion instruments.
- Noise is a sound consisting of a random mix of frequencies.

Using Key Terms

1. Use each of the following terms in a separate sentence: *sound quality* and *noise*.

Understanding Key Ideas

2. What interaction of sound waves determines sound quality?
   a. reflection   **c.** pitch
   b. diffraction   **d.** interference

3. Why do different instruments have different sound qualities?

Critical Thinking

4. **Making Comparisons** What do string instruments and wind instruments have in common in how they produce sound?

5. **Identifying Bias** Someone says that the music you are listening to is "just noise." Does the person mean that the music is a random mix of frequencies? Explain your answer.

Interpreting Graphics

6. Look at the oscilloscope screen below. Do you think the sound represented by the wave on the screen is noise or music? Explain your answer.

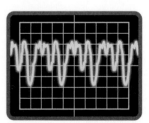

For a variety of links related to this chapter, go to www.scilinks.org

Topic: Sound Quality
SciLinks code: HSM1427

Skills Practice Lab

Easy Listening

Pitch describes how low or high a sound is. A sound's pitch is related to its frequency—the number of waves per second. Frequency is measured in hertz (Hz), where 1 Hz equals 1 wave per second. Most humans can hear frequencies in the range from 20 Hz to 20,000 Hz. But not everyone detects all pitches equally well at all distances. In this activity, you will collect data to see how well you and your classmates hear different frequencies at different distances.

OBJECTIVES

Measure your classmates' ability to detect different pitches at different distances.

Graph the average class data.

Form a conclusion about how easily pitches of different frequencies are heard at different distances.

MATERIALS

- eraser, hard rubber
- meterstick
- paper, graph
- tuning forks, different frequencies (4)

Ask a Question

1. Do most of the students in your classroom hear low-, mid-, or high-frequency sounds best?

Form a Hypothesis

2. Write a hypothesis that answers the question above. Explain your reasoning.

Test the Hypothesis

3. Choose one member of your group to be the sound maker. The others will be the listeners.

4. Copy the data table below onto another sheet of paper. Be sure to include a column for every listener in your group.

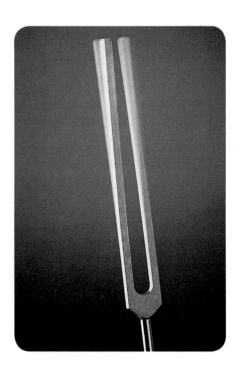

Data Collection Table				
	Distance (m)			
Frequency	Listener 1	Listener 2	Listener 3	Average
1 (____Hz)				
2 (____Hz)		DO NOT WRITE IN BOOK		
3 (____Hz)				
4 (____Hz)				

5. The sound maker will choose one of the tuning forks, and record the frequency of the tuning fork in the data table.

6. The listeners should stand 1 m from the sound maker with their backs turned.

7. The sound maker will create a sound by striking the tip of the tuning fork gently with the eraser.

8. Listeners who hear the sound should take one step away from the sound maker. The listeners who do not hear the sound should stay where they are.

9. Repeat steps 7 and 8 until none of the listeners can hear the sound or the listeners reach the edge of the room.

10. Using the meterstick, the sound maker should measure the distance from his or her position to each of the listeners. All group members should record this data.

11. Repeat steps 5 through 10 with a tuning fork of a different frequency.

12. Continue until all four tuning forks have been tested.

Analyze the Results

1. **Organizing Data** Calculate the average distance for each frequency. Share your group's data with the rest of the class to make a data table for the whole class.

2. **Analyzing Data** Calculate the average distance for each frequency for the class.

3. **Constructing Graphs** Make a graph of the class results, plotting average distance (*y*-axis) versus frequency (*x*-axis).

Draw Conclusions

4. **Drawing Conclusions** Was everyone in the class able to hear all of frequencies equally? (Hint: Was the average distance for each frequency the same?)

5. **Evaluating Data** If the answer to question 4 is no, which frequency had the longest average distance? Which frequency had the shortest final distance?

6. **Analyzing Graphs** Based on your graph, do your results support your hypothesis? Explain your answer.

7. **Evaluating Methods** Do you think your class sample is large enough to confirm your hypothesis for all people of all ages? Explain your answer.

Chapter Review

USING KEY TERMS

Complete each of the following sentences by choosing the correct term from the word bank.

loudness	echoes
pitch	noise
sound quality	

1 The _____ of a sound wave depends on its amplitude.

2 Reflected sound waves are called _____.

3 Two different instruments playing the same note sound different because of _____.

UNDERSTANDING KEY IDEAS

Multiple Choice

4 If a fire engine is traveling toward you, the Doppler effect will cause the siren to sound

a. higher. c. louder.

b. lower. d. softer.

5 Sound travels fastest through

a. a vacuum. c. air.

b. sea water. d. glass.

6 If two sound waves interfere constructively, you will hear

a. a high-pitched sound.

b. a softer sound.

c. a louder sound.

d. no change in sound.

7 You will hear a sonic boom when

a. an object breaks the sound barrier.

b. an object travels at supersonic speeds.

c. a shock wave reaches your ears.

d. the speed of sound is 290 m/s.

8 Resonance can happen when an object vibrates at another object's

a. resonant frequency.

b. fundamental frequency.

c. second overtone frequency.

d. All of the above

9 A technological device that can be used to see sound waves is a(n)

a. sonar. c. ultrasound.

b. oscilloscope. d. amplifier.

Short Answer

10 Describe how the Doppler effect helps a beluga whale determine whether a fish is moving away from it or toward it.

11 How do vibrations cause sound waves?

12 Briefly describe what happens in the different parts of the ear.

Math Skills

13 A submarine that is not moving sends out a sonar sound wave traveling 1,500 m/s, which reflects off a boat back to the submarine. The sonar crew detects the reflected wave 6 s after it was sent out. How far away is the boat from the submarine?

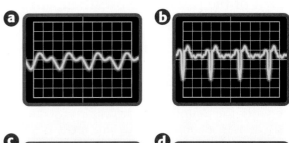

CRITICAL THINKING

14 Concept Mapping Use the following terms to create a concept map: *sound waves, pitch, loudness, decibels, frequency, amplitude, oscilloscope, hertz,* and *interference.*

15 Analyzing Processes An *anechoic chamber* is a room where there is almost no reflection of sound waves. Anechoic chambers are often used to test sound equipment, such as stereos. The walls of such chambers are usually covered with foam triangles. Explain why this design eliminates echoes in the room.

16 Applying Concepts Would the pilot of an airplane breaking the sound barrier hear a sonic boom? Explain why or why not.

17 Forming Hypotheses After working in a factory for a month, a man you know complains about a ringing in his ears. What might be wrong with him? What do you think may have caused his problem? What can you suggest to him to prevent further hearing loss?

INTERPRETING GRAPHICS

Use the oscilloscope screens below to answer the questions that follow:

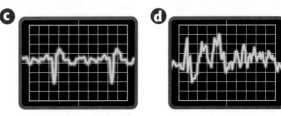

18 Which sound is noise?

19 Which represents the softest sound?

20 Which represents the sound with the lowest pitch?

21 Which two sounds were produced by the same instrument?

Standardized Test Preparation

Read each of the passages below. Then, answer the questions that follow each passage.

Passage 1 Centuries ago, Marco Polo wrote about the booming sand dunes of the Asian desert. He wrote that the booming sands filled the air with the sounds of music, drums, and weapons of war. Booming sands are most often found in the middle of large deserts. They have been discovered all over the world, including the United States. Booming sands make loud, low-pitched sounds when the top layers of sand slip over the layers below, producing vibrations. The sounds have been compared to foghorns, cannon fire, and moaning. The sounds can last from a few seconds to 15 min and can be heard more than 10 km away!

1. Which is a fact in this passage?
   A Marco Polo loved traveling.
   B Booming sands always sound like moaning people.
   C Booming sands are the most interesting thing in Asia.
   D Some booming sands are found in the United States.

2. Which of the following phrases **best** describes booming sands?
   F found in Asia
   G noisy
   H slippery
   I discovered by Marco Polo

3. What causes booming sands?
   A vibrations caused by top layers of sand slipping over layers below
   B battles in the desert
   C animals that live beneath sand dunes
   D There is not enough information to determine the answer.

Passage 2 People who work in the field of architectural acoustics are concerned with controlling sound that travels in a closed space. Their goal is to make rooms and buildings quiet yet suitable for people to enjoy talking and listening to music. One major factor that affects the acoustical quality of a room is the way the room reflects sound waves. Sound waves bounce off surfaces such as doors, ceilings, and walls. Using materials that absorb sound reduces the reflection of sound waves. Materials that have small pockets of air that can trap the sound vibrations and keep them from reflecting are the most sound absorbent. Sound-absorbing floor and ceiling tiles, curtains, and upholstered furniture all help to control the reflection of sound waves.

1. The field of architectural acoustics is concerned with which of the following?
   A making buildings earthquake safe
   B controlling sound in closed spaces
   C designing sound-absorbing materials
   D making buildings as quiet as possible

2. Which of the following is a major factor in the acoustical quality of a room?
   F the size of the room
   G the furnishings in the room
   H the walls of the room
   I the noise level in the room

3. Which of the following materials is **most** likely to absorb sounds the best?
   A materials that have small pockets of air
   B surfaces such as doors, ceilings, and walls
   C materials that keep the room as quiet as possible
   D furniture that is made of wood

Use the pictures of standing waves below to answer the questions that follow.

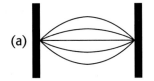

(a)

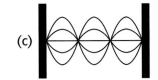

(c)

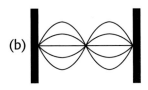

(b)

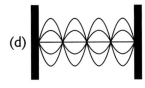

(d)

1. Which of the standing waves has the lowest frequency?

A a

B b

C c

D d

2. Which of the standing waves has the highest frequency?

F a

G b

H c

I d

3. Which of the standing waves represents the first overtone?

A a

B b

C c

D d

4. In which of the following pairs of standing waves is the frequency of the second wave twice the frequency of the first?

F a, b

G a, c

H b, c

I c, d

Read each question below, and choose the best answer.

1. The speed of sound in copper is 3,560 m/s. Which is another way to express this measure?

A 356×10^2 m/s

B 0.356×10^3 m/s

C 3.56×10^3 m/s

D 3.56×10^4 m/s

2. The speed of sound in sea water is 1,522 m/s. How far can a sound wave travel underwater in 10 s?

F 152.2 m

G 1,522 m

H 15,220 m

I 152,220 m

3. Claire likes to go swimming after work. She warms up for 120 s before she begins swimming, and it takes her an average of 55 s to swim one lap. Which equation could be used to find w, the number of seconds it takes for Claire to warm up and swim 15 laps?

A $w = (15 \times 120) + 55$

B $w = (15 \times 55) + 120$

C $w = 120 + 55 + 15$

D $w = (15 \times 55) \times 120$

4. The Vasquez family went bowling. They rented 6 pairs of shoes for $3 a pair and bowled for 2 h at a rate of $8.80/h. Which is the best estimate of the total cost of the shoes and bowling?

F $24

G $30

H $36

I $45

Standardized Test Preparation

Science in Action

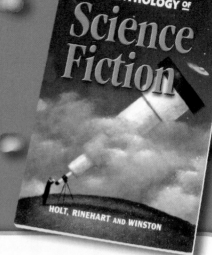

HOLT ANTHOLOGY OF
Science Fiction

HOLT, RINEHART AND WINSTON

Scientific Discoveries

Jurassic Bark

Imagine you suddenly hear a loud honking sound, such as a trombone or a tuba. "Must be band tryouts," you think. You turn to find the noise and find yourself face to face with a 10 m long, 2,800 kg dinosaur with a huge tubular crest on its snout. Do you run? No—your musical friend, *Parasaurolophus,* is a vegetarian. In 1995, an almost-complete fossil skull of an adult *Parasaurolophus* was found in New Mexico. Scientists studied the noise-making qualities of *Parasaurolophus*'s crest and found that it contained many internal tubes and chambers.

Math ACTiViTY

Imagine that a standing wave with a frequency of 80 Hz is made inside the crest of a *Parasaurolophus*. What would be the frequency of the first overtone of this standing wave? the second? the third?

Science Fiction

"Ear" by Jane Yolen

Jily and her friends, Sanya and Feeny, live in a time not too far in the future. It is a time when everyone's hearing is damaged. People communicate using sign language—unless they put on their Ear. Then, the whole world is filled with sounds.

Jily and her friends visit a club called The Low Down. It is too quiet for Jily's tastes, and she wants to leave. But Sanya is dancing by herself, even though there is no music. When Jily finds Feeny, they notice some Earless kids their own age. Earless people never go to clubs, and Jily finds their presence offensive. But Feeny is intrigued.

Everyone is given an Ear at the age of 12 but has to give it up at the age of 30. Why would these kids want to go out without their Ears before the age of 30? Jily thinks the idea is ridiculous and doesn't stick around to find out the answer to such a question. But it is an answer that will change her life by the end of the next day.

Language Arts ACTiViTY

WRITING SKILL Read "Ear," by Jane Yolen, in the *Holt Anthology of Science Fiction.* Write a one-page report that discusses how the story made you think about the importance of hearing in your everyday life.

Adam Dudley

Sound Engineer Adam Dudley uses the science of sound waves every day at his job. He is the audio supervisor for the Performing Arts Center of the University of Texas at Austin. Dudley oversees sound design and technical support for campus performance spaces, including an auditorium that seats over 3,000 people.

To stage a successful concert, Dudley takes many factors into account. The size and shape of the room help determine how many speakers to use and where to place them. It is a challenge to make sure people seated in the back row can hear well enough and also to make sure that the people up front aren't going deaf from the high volume.

Adam Dudley loves his job—he enjoys working with people and technology and prefers not to wear a coat and tie. Although he is invisible to the audience, his work backstage is as crucial as the musicians and actors on stage to the success of the events.

Social Studies ACTIVITY

Research the ways in which concert halls were designed before the use of electric amplification. Make a model or diorama, and present it to the class, explaining the acoustical factors involved in the design.

go.hrw.com

To learn more about these Science in Action topics, visit **go.hrw.com** and type in the keyword **HP5SNDF.**

Current Science

Check out Current Science® articles related to this chapter by visiting go.hrw.com. Just type in the keyword **HP5CS21.**

3

The Nature of Light

The Big Idea

Light is an electromagnetic wave. Electromagnetic waves interact in predictable ways.

About the PHOTO

What kind of alien life lives on this planet? Actually, this isn't a planet at all. It's an ordinary soap bubble! The brightly colored swirls on this bubble are reflections of light. Light waves combine through interference so that you see different colors on this soap bubble.

PRE-READING ACTIVITY

FOLDNOTES **Booklet** Before you read the chapter, create the FoldNote entitled "Booklet" described in the **Study Skills** section of the Appendix. Label each page of the booklet with a main idea from the chapter. As you read the chapter, write what you learn about each main idea on the appropriate page of the booklet.

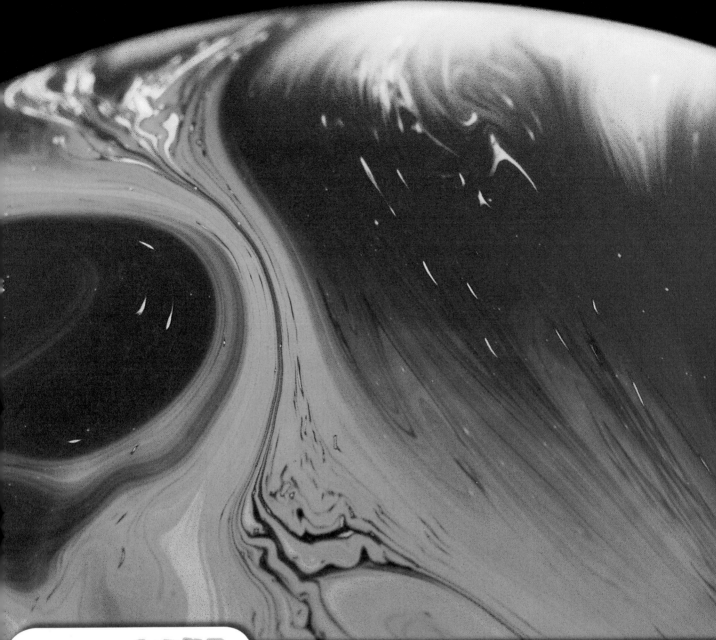

START-UP ACTIVITY

Colors of Light

Is white light really white? In this activity, you will use a spectroscope to answer that question.

Procedure

1. Your teacher will give you a **spectroscope** or instructions for making one.

2. Turn on an **incandescent light bulb.** Look at the light bulb through your spectroscope. Write a description of what you see.

3. Repeat step 2, looking at a **fluorescent light.** Again, describe what you see.

Analysis

1. Compare what you saw with the incandescent light bulb with what you saw with the fluorescent light bulb.

2. Both kinds of bulbs produce white light. What did you learn about white light by using the spectroscope?

3. Light from a flame is yellowish but is similar to white light. What do you think you would see if you used a spectroscope to look at light from a flame?

SECTION 1

What Is Light?

You can see light. It's everywhere! Light comes from the sun and from other sources, such as light bulbs. But what exactly is light?

Scientists are still studying light to learn more about it. A lot has already been discovered about light, as you will soon find out. Read on, and be enlightened!

electromagnetic wave a wave that consists of electric and magnetic fields that vibrate at right angles to each other

Light: An Electromagnetic Wave

Light is a type of energy that travels as a wave. But light is different from other kinds of waves. Other kinds of waves, like sound waves and water waves, must travel through matter. Light does not require matter through which to travel. Light is an electromagnetic wave (EM wave). An **electromagnetic wave** is a wave that can travel through empty space or matter and consists of changing electric and magnetic fields.

Fields exist around certain objects and can exert a force on another object without touching that object. For example, Earth is a source of a gravitational field. This field pulls you and all things toward Earth. But keep in mind that this field, like all fields, is not made of matter.

Figure 1 shows a diagram of an electromagnetic wave. Notice that the electric and magnetic fields are at right angles—or are *perpendicular*—to each other. These fields are also perpendicular to the direction of the wave motion.

Figure 1 *Electromagnetic waves are made of vibrating electric and magnetic fields.*

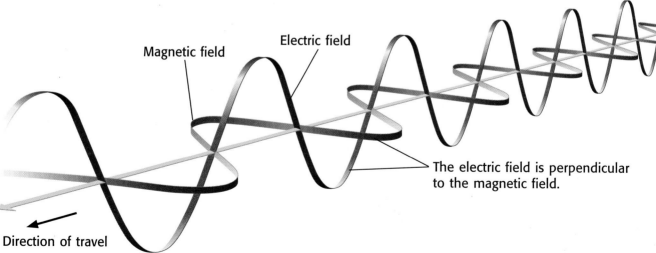

Magnetic field

Electric field

The electric field is perpendicular to the magnetic field.

Direction of travel

Figure 2 *The hair on the girl's head stands up because of an electric field and the iron filings form arcs around the magnet because of a magnetic field.*

Electric and Magnetic Fields

Electromagnetic waves are changing electric and magnetic fields. But what are electric and magnetic fields? An *electric field* surrounds every charged object. The electric field around a charged object pulls oppositely charged objects toward it and repels like-charged objects. You can see the effect of electric fields whenever you see objects stuck together by static electricity. **Figure 2** shows another effect of an electric field.

A *magnetic field* surrounds every magnet. Because of magnetic fields, paper clips and iron filings are pulled toward magnets. You can feel the effect of magnetic fields when you hold two magnets close together. The iron filings around the magnet in **Figure 2** form arcs in the presence of the magnet's magnetic field.

✓ Reading Check **Where can electric fields be found?**
(See the Appendix for answers to Reading Checks.)

How EM Waves Are Produced

An EM wave can be produced by the vibration of an electrically charged particle. When the particle vibrates, or moves back and forth, the electric field around it also vibrates. When the electric field starts vibrating, a vibrating magnetic field is created. The vibration of an electric field and a magnetic field together produces an EM wave that carries energy released by the original vibration of the particle. The transfer of energy as electromagnetic waves is called **radiation.**

CONNECTION TO Social Studies

WRITING SKILL **The Particle Model of Light**
Thinking of light as being an electromagnetic wave can explain many properties of light. But some properties of light can be explained only by using a particle model of light. In the particle model of light, light is thought of as a stream of particles called *photons*. Research the history of the particle model of light. Write a one-page paper on what you learn.

radiation transfer of energy as electromagnetic waves

Figure 3 *Thunder and lightning are produced at the same time. But you usually see lightning before you hear thunder, because light travels much faster than sound.*

The Speed of Light

Scientists have yet to discover anything that travels faster than light. In the near vacuum of space, the speed of light is about 300,000,000 m/s, or 300,000 km/s. Light travels slightly slower in air, glass, and other types of matter. (Keep in mind that even though electromagnetic waves do not need to travel through matter, they can travel through many substances.)

Believe it or not, light can travel about 880,000 times faster than sound! This fact explains the phenomenon described in **Figure 3.** If you could run at the speed of light, you could travel around Earth 7.5 times in 1 s.

✓ Reading Check How does the speed of light compare with the speed of sound?

MATH FOCUS

How Fast Is Light? The distance from Earth to the moon is 384,000 km. Calculate the time it takes for light to travel that distance.

Step 1: Write the equation for speed.

$$speed = \frac{distance}{time}$$

Step 2: Rearrange the equation by multiplying by time and dividing by speed.

$$time = \frac{distance}{speed}$$

Step 3: Replace *distance* and *speed* with the values given in the problem, and solve.

$$time = \frac{384,000 \text{ km}}{300,000 \text{ km/s}}$$
$$time = 1.28 \text{ s}$$

Now It's Your Turn

1. The distance from the sun to Venus is 108,000,000 km. Calculate the time it takes for light to travel that distance.

Light from the Sun

Even though light travels quickly, it takes about 8.3 min for light to travel from the sun to Earth. It takes this much time because Earth is 150,000,000 km away from the sun.

The EM waves from the sun are the major source of energy on Earth. For example, plants use photosynthesis to store energy from the sun. And animals use and store energy by eating plants or by eating other animals that eat plants. Even fossil fuels, such as coal and oil, store energy from the sun. Fossil fuels are formed from the remains of plants and animals that lived millions of years ago.

Although Earth receives a large amount of energy from the sun, only a very small part of the total energy given off by the sun reaches Earth. Look at **Figure 4.** The sun gives off energy as EM waves in all directions. Most of this energy travels away in space.

Figure 4 *Only a small amount of the sun's energy reaches the planets in the solar system.*

SECTION Review

Summary

- Light is an electromagnetic (EM) wave. An EM wave is a wave that consists of changing electric and magnetic fields. EM waves require no matter through which to travel.

- EM waves can be produced by the vibration of charged particles.

- The speed of light in a vacuum is about 300,000,000 m/s.

- EM waves from the sun are the major source of energy for Earth.

Using Key Terms

1. Use the following terms in the same sentence: *electromagnetic wave* and *radiation*.

Understanding Key Ideas

2. Electromagnetic waves are different from other types of waves because they can travel through

   a. air.   **c.** space.

   b. glass.   **d.** steel.

3. Describe light in terms of electromagnetic waves.

4. Why is light from the sun important?

5. How can electromagnetic waves be produced?

Math Skills

6. The distance from the sun to Jupiter is 778,000,000 km. How long does it take for light from the sun to reach Jupiter?

Critical Thinking

7. **Making Inferences** Why is it important that EM waves can travel through empty space?

8. **Making Comparisons** How does the amount of energy produced by the sun compare with the amount of energy that reaches Earth from the sun?

9. **Applying Concepts** Explain why the energy produced by burning wood in a campfire is energy from the sun.

SC*i*LINKS.

NSTA
Developed and maintained by the National Science Teachers Association

For a variety of links related to this chapter, go to www.scilinks.org

Topic: Light Energy
SciLinks code: HSM0880

The Electromagnetic Spectrum

When you look around, you can see things that reflect light to your eyes. But a bee might see the same things differently. Bees can see a kind of light—called **ultraviolet light**—that you can't see!

It might seem odd to call something you can't see *light*. The light you are most familiar with is called *visible light*. Ultraviolet light is similar to visible light. Both are kinds of electromagnetic (EM) waves. In this section, you will learn about many kinds of EM waves, including X rays, radio waves, and microwaves.

Characteristics of EM Waves

All EM waves travel at the same speed in a vacuum— 300,000 km/s. How is this possible? The speed of a wave is found by multiplying its wavelength by its frequency. So, EM waves having different wavelengths can travel at the same speed as long as their frequencies are also different. The entire range of EM waves is called the **electromagnetic spectrum.** The electromagnetic spectrum is shown in **Figure 1.** The electromagnetic spectrum is divided into regions according to the length of the waves. There is no sharp division between one kind of wave and the next. Some kinds even have overlapping ranges.

✓ Reading Check How is the speed of a wave determined? (*See the Appendix for answers to Reading Checks.*)

Figure 1 **The Electromagnetic Spectrum**

The electromagnetic spectrum is arranged from long to short wavelength or from low to high frequency.

Radio waves

All radio and television stations broadcast radio waves.

Microwaves

Despite their name, microwaves are not the shortest EM waves.

Infrared

Infrared means "below red."

Radio Waves

Radio waves cover a wide range of waves in the EM spectrum. Radio waves have some of the longest wavelengths and the lowest frequencies of all EM waves. In fact, radio waves are any EM waves that have wavelengths longer than 30 cm. Radio waves are used for broadcasting radio signals.

Broadcasting Radio Signals

Figure 2 shows how radio signals are broadcast. Radio stations encode sound information into radio waves by varying either the waves' amplitude or their frequency. Changing amplitude or frequency is called *modulation* (MAHJ uh LAY shuhn). You probably know that there are AM radio stations and FM radio stations. The abbreviation *AM* stands for "amplitude modulation," and the abbreviation *FM* stands for "frequency modulation."

Comparing AM and FM Radio Waves

AM radio waves are different from FM radio waves. For example, AM radio waves have longer wavelengths than FM radio waves do. And AM radio waves can bounce off the atmosphere and thus can travel farther than FM radio waves. But FM radio waves are less affected by electrical noise than AM radio waves are. So, music broadcast from FM stations sounds better than music broadcast from AM stations.

❶ A radio station converts sound into an electric current. The current produces radio waves that are sent out in all directions by the antenna.

❷ A radio receives radio waves and then converts them into an electric current, which is then converted to sound.

Figure 2 *Radio waves cannot be heard, but they can carry energy that can be converted into sound.*

electromagnetic spectrum all of the frequencies or wavelengths of electromagnetic radiation

---

Decreasing wavelength/Increasing frequency →

Visible light
Visible light contains all of the colors that you can see.

Ultraviolet
Ultraviolet means "beyond violet."

X rays
X rays were discovered in 1895.

Gamma rays
Gamma rays are produced by some nuclear reactions.

Radio Waves and Television

Television signals are also carried by radio waves. Most television stations broadcast radio waves that have shorter wavelengths and higher frequencies than those broadcast by radio stations. Like radio signals, television signals are broadcast using amplitude modulation and frequency modulation. Television stations use frequency-modulated waves to carry sound and amplitude-modulated waves to carry pictures.

Some waves carrying television signals are transmitted to artificial satellites orbiting Earth. The waves are amplified and sent to ground antennas. They then travel through cables to televisions in homes. Cable television works by this process.

✓ Reading Check Which EM waves can carry television signals?

Microwaves

Microwaves have shorter wavelengths and higher frequencies than radio waves do. Microwaves have wavelengths between 1 mm and 30 cm. You are probably familiar with microwaves—they are created in a microwave oven, such as the one shown in **Figure 3.**

Microwaves and Communication

Like radio waves, microwaves are used to send information over long distances. For example, cellular phones send and receive signals using microwaves. And signals sent between Earth and artificial satellites in space are also carried by microwaves.

Figure 3 How a Microwave Oven Works

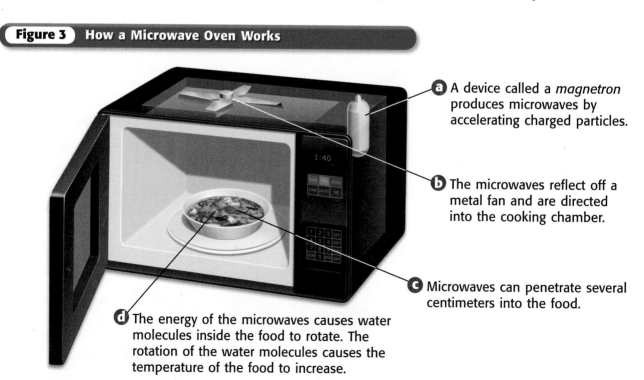

a A device called a *magnetron* produces microwaves by accelerating charged particles.

b The microwaves reflect off a metal fan and are directed into the cooking chamber.

c Microwaves can penetrate several centimeters into the food.

d The energy of the microwaves causes water molecules inside the food to rotate. The rotation of the water molecules causes the temperature of the food to increase.

Figure 4 *Police officers use radar to detect cars going faster than the speed limit.*

Radar

Microwaves are also used in radar. *Radar* (**ra**dio **d**etection **a**nd **r**anging) is used to detect the speed and location of objects. The police officer in **Figure 4** is using radar to check the speed of a car. The radar gun sends out microwaves that reflect off the car and return to the gun. The reflected waves are used to calculate the speed of the car. Radar is also used to watch the movement of airplanes and to help ships navigate at night.

Infrared Waves

Infrared waves have shorter wavelengths and higher frequencies than microwaves do. The wavelengths of infrared waves vary between 700 nanometers and 1 mm. A nanometer (nm) is equal to 0.000000001 m.

On a sunny day, you may be warmed by infrared waves from the sun. Your skin absorbs infrared waves striking your body. The energy of the waves causes the particles in your skin to vibrate more, and you feel an increase in temperature. The sun is not the only source of infrared waves. Almost all things give off infrared waves, including buildings, trees, and you! The amount of infrared waves an object gives off depends on the object's temperature. Warmer objects give off more infrared waves than cooler objects do.

You can't see infrared waves, but some devices can detect infrared waves. For example, infrared binoculars change infrared waves into light you can see. Such binoculars can be used to watch animals at night. **Figure 5** shows a photo taken with film that is sensitive to infrared waves.

Figure 5 *In this photograph, brighter colors indicate higher temperatures.*

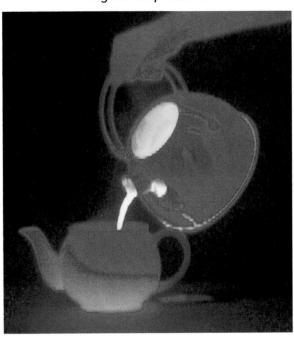

Figure 6 *Water droplets can separate white light into visible light of different wavelengths. As a result, you see all the colors of visible light in a rainbow.*

Making a Rainbow

On a sunny day, ask an adult to use a hose or a spray bottle to make a mist of water outside. Move around until you see a rainbow in the water mist. Draw a diagram showing the positions of the water mist, the sun, the rainbow, and yourself.

Visible Light

Visible light is the very narrow range of wavelengths and frequencies in the electromagnetic spectrum that humans can see. Visible light waves have shorter wavelengths and higher frequencies than infrared waves do. Visible light waves have wavelengths between 400 nm and 700 nm.

Visible Light from the Sun

Some of the energy that reaches Earth from the sun is visible light. The visible light from the sun is white light. *White light* is visible light of all wavelengths combined. Light from lamps in your home as well as from the fluorescent bulbs in your school is also white light.

✓ Reading Check What is white light?

Colors of Light

Humans see the different wavelengths of visible light as different colors, as shown in **Figure 6.** The longest wavelengths are seen as red light. The shortest wavelengths are seen as violet light.

The range of colors is called the *visible spectrum*. You can see the visible spectrum in **Figure 7.** When you list the colors, you might use the imaginary name *ROY G. BiV* to help you remember their order. The capital letters in Roy's name represent the first letter of each color of visible light: **r**ed, **o**range, **y**ellow, **g**reen, **b**lue, and **v**iolet. What about the *i* in Roy's last name? You can think of *i* as standing for the color indigo. Indigo is a dark blue color.

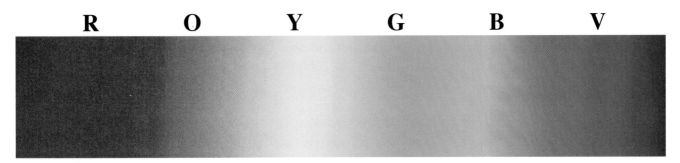

R   O   Y   G   B   V

Figure 7 *The visible spectrum contains all colors of light.*

Ultraviolet Light

Ultraviolet light (UV light) is another type of electromagnetic wave produced by the sun. Ultraviolet waves have shorter wavelengths and higher frequencies than visible light does. The wavelengths of ultraviolet light waves vary between 60 nm and 400 nm. Ultraviolet light affects your body in both bad and good ways.

Reading Check How do ultraviolet light waves compare with visible light waves?

Bad Effects

On the bad side, too much ultraviolet light can cause sunburn, as you can see in **Figure 8.** Too much ultraviolet light can also cause skin cancer, wrinkles, and damage to the eyes. Luckily, much of the ultraviolet light from the sun does not reach Earth's surface. But you should still protect yourself against the ultraviolet light that does reach you. To do so, you should use sunscreen with a high SPF (sun protection factor). You should also wear sunglasses that block out UV light to protect your eyes. Hats, long-sleeved shirts, and long pants can protect you, too. You need this protection even on overcast days because UV light can travel through clouds.

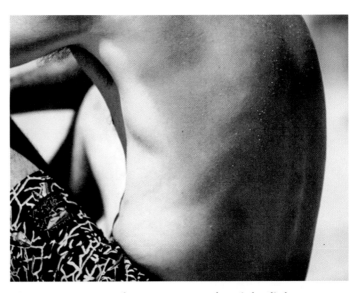

Figure 8 *Too much exposure to ultraviolet light can lead to a painful sunburn. Using sunscreen will help protect your skin.*

Good Effects

On the good side, ultraviolet waves produced by ultraviolet lamps are used to kill bacteria on food and surgical tools. In addition, small amounts of ultraviolet light are beneficial to your body. When exposed to ultraviolet light, skin cells produce vitamin D. This vitamin allows the intestines to absorb calcium. Without calcium, your teeth and bones would be very weak.

X Rays and Gamma Rays

X rays and gamma rays have some of the shortest wavelengths and highest frequencies of all EM waves.

X Rays

X rays have wavelengths between 0.001 nm and 60 nm. They can pass through many materials. This characteristic makes X rays useful in the medical field, as shown in **Figure 9.** But too much exposure to X rays can also damage or kill living cells. A patient getting an X ray may wear special aprons to protect parts of the body that do not need X-ray exposure. These aprons are lined with lead because X rays cannot pass through lead.

X-ray machines are also used as security devices in airports and other public buildings. The machines allow security officers to see inside bags and other containers without opening the containers.

✓ **Reading Check** How are patients protected from X rays?

Gamma Rays

Gamma rays are EM waves that have wavelengths shorter than 0.1 nm. They can penetrate most materials very easily. Gamma rays are used to treat some forms of cancer. Doctors focus the rays on tumors inside the body to kill the cancer cells. This treatment often has good effects, but it can have bad side effects because some healthy cells may also be killed.

Gamma rays are also used to kill harmful bacteria in foods, such as meat and fresh fruits. The gamma rays do not harm the treated food and do not stay in the food. So, food that has been treated with gamma rays is safe for you to eat.

CONNECTION TO Astronomy

Gamma Ray Spectrometer
In 2001, NASA put an artificial satellite called the *2001 Mars Odyssey* in orbit around Mars. The *Odyssey* is carrying a gamma ray spectrometer. A *spectrometer* is a device used to detect certain kinds of EM waves. The gamma ray spectrometer on the *Odyssey* was used to look for water and several chemical elements on Mars. Scientists hope to use this information to learn about the geology of Mars. Research the characteristics of Mars and Earth. In your **science journal,** make a chart comparing Mars and Earth.

ACTIVITY

Figure 9 How a Bone Is X Rayed

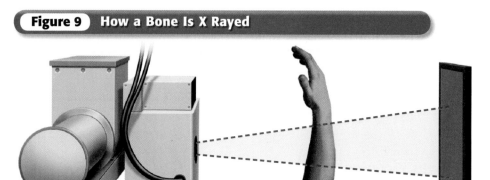
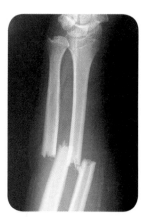

❶ X rays travel easily through skin and muscle but are absorbed by bones.

❷ The X rays that are not absorbed strike the film.

❸ Bright areas appear on the film where X rays are absorbed by the bones.

Summary

- All electromagnetic (EM) waves travel at the speed of light. EM waves differ only by wavelength and frequency.
- The entire range of EM waves is called the *electromagnetic spectrum*.
- Radio waves are used for communication.
- Microwaves are used in cooking and in radar.
- The absorption of infrared waves is felt as an increase in temperature.

- Visible light is the narrow range of wavelengths that humans can see. Different wavelengths are seen as different colors.
- Ultraviolet light is useful for killing bacteria and for producing vitamin D in the body. Overexposure to ultraviolet light can cause health problems.
- X rays and gamma rays are EM waves that are often used in medicine. Overexposure to these kinds of rays can damage or kill living cells.

Using Key Terms

1. In your own words, write a definition for the term *electromagnetic spectrum*.

Understanding Key Ideas

2. Which of the following electromagnetic waves are produced by the sun?
   - **a.** infrared waves
   - **b.** visible light
   - **c.** ultraviolet light
   - **d.** All of the above

3. How do the different kinds of EM waves differ from each other?

4. Describe two ways of transmitting information using radio waves.

5. Explain why ultraviolet light, X rays, and gamma rays can be both helpful and harmful.

6. What are two common uses for microwaves?

7. What is white light? What are two sources of white light?

8. What is the visible spectrum?

Critical Thinking

9. **Applying Concepts** Describe how three different kinds of electromagnetic waves have been useful to you today.

10. **Making Comparisons** Compare the wavelengths of infrared waves, ultraviolet light, and visible light.

Interpreting Graphics

The waves in the diagram below represent two different kinds of EM waves. Use the diagram below to answer the questions that follow.

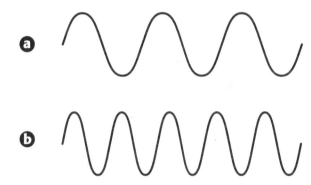

11. Which wave has the longest wavelength?

12. Suppose that one of the waves represents a microwave and one of the waves represents a radio wave. Which wave represents the microwave?

SCiLINKS®

Developed and maintained by the
National Science Teachers Association

For a variety of links related to this chapter, go to www.scilinks.org

Topic: Electromagnetic Spectrum
SciLinks code: HSM0482

Interactions of Light Waves

Have you ever seen a cat's eyes glow in the dark when light shines on them? Cats have a special layer of cells in the back of their eyes that reflects light.

This layer helps the cat see better by giving the eyes another chance to detect the light. Reflection is one interaction of electromagnetic waves. Because we can see visible light, it is easier to explain all wave interactions by using visible light.

Reflection

Reflection happens when light waves bounce off an object. Light reflects off objects all around you. When you look in a mirror, you are seeing light that has been reflected twice—first from you and then from the mirror. If light is reflecting off everything around you, why can't you see your image on a wall? To answer this question, you must learn the law of reflection.

The Law of Reflection

Light reflects off surfaces the same way that a ball bounces off the ground. If you throw the ball straight down against a smooth surface, it will bounce straight up. If you bounce it at an angle, it will bounce away at an angle. The *law of reflection* states that the angle of incidence is equal to the angle of reflection. *Incidence* is the arrival of a beam of light at a surface. **Figure 1** shows this law.

✓ **Reading Check** What is the law of reflection? (*See the Appendix for answers to Reading Checks.*)

What You Will Learn

● Describe how reflection allows you to see things.
● Describe absorption and scattering.
● Explain how refraction can create optical illusions and separate white light into colors.
● Explain the relationship between diffraction and wavelength.
● Compare constructive and destructive interference of light.

Vocabulary

reflection
absorption
scattering

refraction
diffraction
interference

READING STRATEGY

Reading Organizer As you read this section, make a concept map by using the terms above.

Figure 1   The Law of Reflection

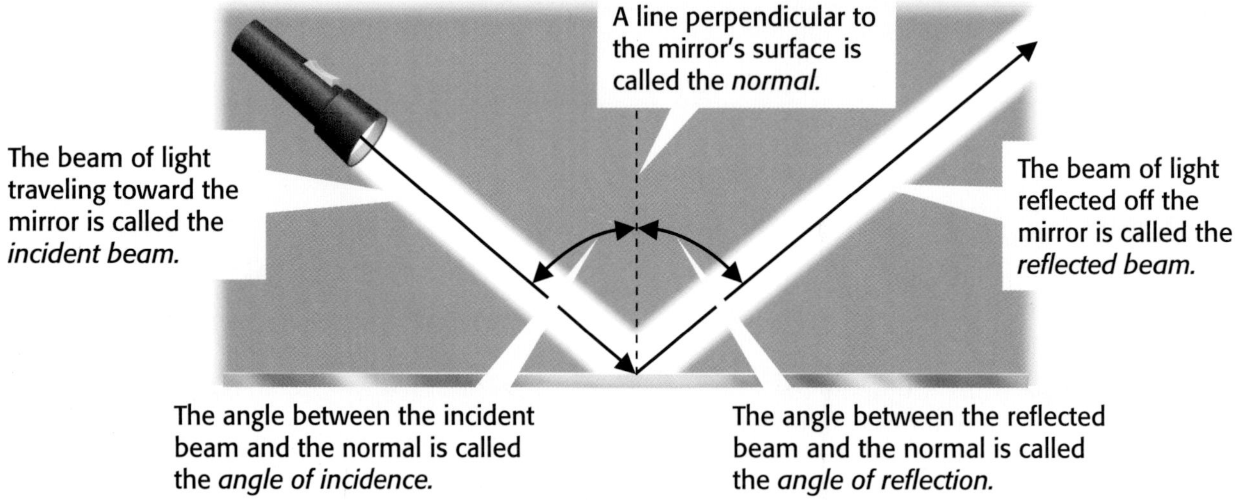

A line perpendicular to the mirror's surface is called the *normal.*

The beam of light traveling toward the mirror is called the *incident beam.*

The beam of light reflected off the mirror is called the *reflected beam.*

The angle between the incident beam and the normal is called the *angle of incidence.*

The angle between the reflected beam and the normal is called the *angle of reflection.*

Figure 2 Regular Reflection Vs. Diffuse Reflection

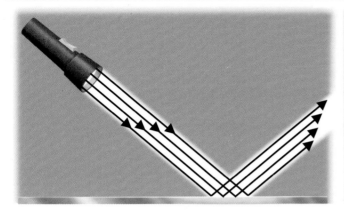 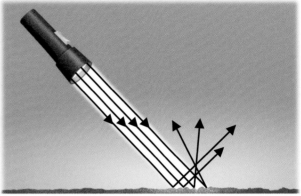

Regular reflection occurs when light beams are reflected at the same angle. When your eye detects the reflected beams, you can see a reflection on the surface.

Diffuse reflection occurs when light beams reflect at many different angles. You can't see a reflection because not all of the reflected light is directed toward your eyes.

Types of Reflection

So, why can you see your image in a mirror but not in a wall? The answer has to do with the differences between the two surfaces. A mirror's surface is very smooth. Thus, light beams reflect off all points of the mirror at the same angle. This kind of reflection is called *regular reflection*. A wall's surface is slightly rough. Light beams will hit the wall's surface and reflect at many different angles. This kind of reflection is called *diffuse reflection*. **Figure 2** shows the difference between the two kinds of reflection.

reflection the bouncing back of a ray of light, sound, or heat when the ray hits a surface that it does not go through

Light Source or Reflection?

If you look at a TV set in a bright room, you see the cabinet around the TV and the image on the screen. But if you look at the same TV in the dark, you see only the image on the screen. The difference is that the screen is a light source, but the cabinet around the TV is not.

You can see a light source even in the dark because its light passes directly into your eyes. The tail of the firefly in **Figure 3** is a light source. Flames, light bulbs, and the sun are also light sources. Objects that produce visible light are called *luminous* (LOO muh nuhs).

Most things around you are not light sources. But you can still see them because light from light sources reflects off the objects and then travels to your eyes. A visible object that is not a light source is *illuminated*.

Figure 3 *You can see the tail of this firefly because it is luminous. But you see its body because it is illuminated.*

✓ **Reading Check** List four different light sources.

Moonlight? Sometimes, the moon shines so brightly that you might think there is a lot of "moonlight." But did you know that moonlight is actually sunlight? The moon does not give off light. You can see the moon because it is illuminated by light from the sun. You see different phases of the moon because light from the sun shines only on the part of the moon that faces the sun. Make a poster that shows the different phases of the moon.

ACTIVITY

Absorption and Scattering

absorption in optics, the transfer of light energy to particles of matter

scattering an interaction of light with matter that causes light to change its energy, direction of motion, or both

Have you noticed that when you use a flashlight, the light shining on things closer to you appears brighter than the light shining on things farther away? The light is less bright the farther it travels from the flashlight. The light is weaker partly because the beam spreads out and partly because of absorption and scattering.

Absorption of Light

The transfer of energy carried by light waves to particles of matter is called **absorption.** When a beam of light shines through the air, particles in the air absorb some of the energy from the light. As a result, the beam of light becomes dim. The farther the light travels from its source, the more it is absorbed by particles, and the dimmer it becomes.

Scattering of Light

Scattering is an interaction of light with matter that causes light to change direction. Light scatters in all directions after colliding with particles of matter. Light from the ship shown in **Figure 4** is scattered out of the beam by air particles. This scattered light allows you to see things that are outside the beam. But, because light is scattered out of the beam, the beam becomes dimmer.

Scattering makes the sky blue. Light with shorter wavelengths is scattered more than light with longer wavelengths. Sunlight is made up of many different colors of light, but blue light (which has a very short wavelength) is scattered more than any other color. So, when you look at the sky, you see a background of blue light.

Figure 4 *A beam of light becomes dimmer partly because of scattering.*

✓ *Reading Check* **Why can you see things outside a beam of light?**

Scattering Milk

1. Fill a **2 L clear plastic bottle** with **water.**

2. Turn the lights off, and shine a **flashlight** through the water. Look at the water from all sides of the bottle. Write a description of what you see.

3. Add **3 drops of milk** to the water, and shake the bottle to mix it up.

4. Repeat step 2. Describe any color changes. If you don't see any, add more milk until you do.

5. How is the water-and-milk mixture like air particles in the atmosphere? Explain your answer.

Refraction

Imagine that you and a friend are at a lake. Your friend wades into the water. You look at her, and her feet appear to have separated from her legs! What has happened? You know her feet did not fall off, so how can you explain what you see? The answer has to do with refraction.

Refraction and Material

Refraction is the bending of a wave as it passes at an angle from one substance, or material, to another. **Figure 5** shows a beam of light refracting twice. Refraction of light waves occurs because the speed of light varies depending on the material through which the waves are traveling. In a vacuum, light travels at 300,000 km/s, but it travels more slowly through matter. When a wave enters a new material at an angle, the part of the wave that enters first begins traveling at a different speed from that of the rest of the wave.

refraction the bending of a wave as the wave passes between two substances in which the speed of the wave differs

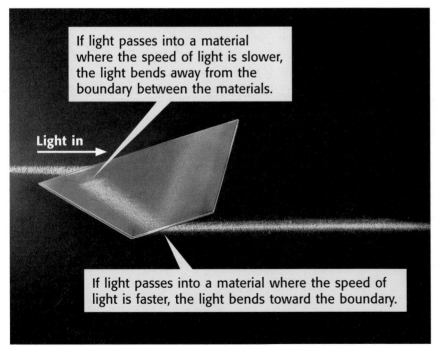

If light passes into a material where the speed of light is slower, the light bends away from the boundary between the materials.

Light in

If light passes into a material where the speed of light is faster, the light bends toward the boundary.

Figure 5 *Light travels more slowly through glass than it does through air. So, light refracts as it passes at an angle from air to glass or from glass to air. Notice that the light is also reflected inside the prism.*

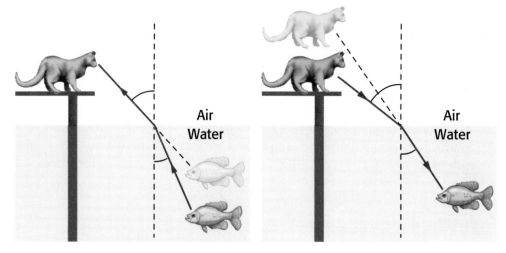

Figure 6 *Because of refraction, the cat and the fish see optical illusions. To the cat, the fish appears closer than it really is. To the fish, the cat appears farther away than it actually is.*

Air
Water

Air
Water

Refraction and Optical Illusions

Usually, when you look at an object, the light reflecting off the object travels in a straight line from the object to your eye. Your brain always interprets light as traveling in straight lines. But when you look at an object that is underwater, the light reflecting off the object does not travel in a straight line. Instead, it refracts. **Figure 6** shows how refraction creates an optical illusion. This kind of illusion causes a person's feet to appear separated from the legs when the person is wading.

Refraction and Color Separation

White light is composed of all the wavelengths of visible light. The different wavelengths of visible light are seen by humans as different colors. When white light is refracted, the amount that the light bends depends on its wavelength. Waves with short wavelengths bend more than waves with long wavelengths. As shown in **Figure 7,** white light can be separated into different colors during refraction. Color separation by refraction is responsible for the formation of rainbows. Rainbows are created when sunlight is refracted by water droplets.

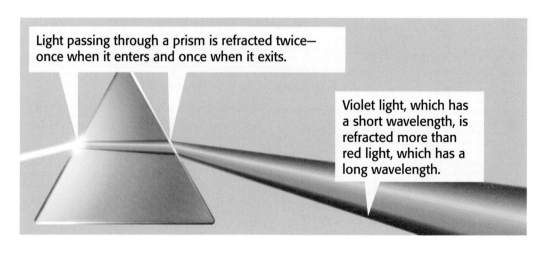

Light passing through a prism is refracted twice—once when it enters and once when it exits.

Figure 7 *A prism is a piece of glass that separates white light into the colors of visible light by refraction.*

Violet light, which has a short wavelength, is refracted more than red light, which has a long wavelength.

Quick Lab

Refraction Rainbow

1. **Tape** a **piece of construction paper** over the end of a **flashlight.** Use **scissors** to cut a slit in the paper.

2. Turn on the flashlight, and lay it on a table. Place a **prism** on end in the beam of light.

3. Slowly rotate the prism until you can see a rainbow on the surface of the table. Draw a diagram of the light beam, the prism, and the rainbow.

Diffraction

Refraction isn't the only way light waves are bent. **Diffraction** is the bending of waves around barriers or through openings. The amount a wave diffracts depends on its wavelength and the size of the barrier or the opening. The greatest amount of diffraction occurs when the barrier or opening is the same size or smaller than the wavelength.

diffraction a change in the direction of a wave when the wave finds an obstacle or an edge, such as an opening

✔ **Reading Check** The amount a wave diffracts depends on what two things?

Diffraction and Wavelength

The wavelength of visible light is very small—about 100 times thinner than a human hair! So, a light wave cannot bend very much by diffraction unless it passes through a narrow opening, around sharp edges, or around a small barrier, as shown in **Figure 8.**

Light waves cannot diffract very much around large obstacles, such as buildings. Thus, you can't see around corners. But light waves always diffract a small amount. You can observe light waves diffracting if you examine the edges of a shadow. Diffraction causes the edges of shadows to be blurry.

Figure 8 *This diffraction pattern is made by light of a single wavelength shining around the edges of a very tiny disk.*

Interference

interference the combination of two or more waves that results in a single wave

Interference is a wave interaction that happens when two or more waves overlap. Overlapping waves can combine by constructive or destructive interference.

Constructive Interference

When waves combine by *constructive interference,* the resulting wave has a greater amplitude, or height, than the individual waves had. Constructive interference of light waves can be seen when light of one wavelength shines through two small slits onto a screen. The light on the screen will appear as a series of alternating bright and dark bands, as shown in **Figure 9.** The bright bands result from light waves combining through constructive interference.

✓ **Reading Check** What is constructive interference?

Destructive Interference

When waves combine by *destructive interference,* the resulting wave has a smaller amplitude than the individual waves had. So, when light waves interfere destructively, the result will be dimmer light. Destructive interference forms the dark bands seen in **Figure 9.**

You do not see constructive or destructive interference of white light. To understand why, remember that white light is composed of waves with many different wavelengths. The waves rarely line up to combine in total destructive interference.

INTERNET ACTIVITY

For another activity related to this chapter, go to **go.hrw.com** and type in the keyword **HP5LGTW.**

Figure 9 Constructive and Destructive Interference

1 Red light of one wavelength passes between two tiny slits.

2 The light waves diffract as they pass through the tiny slits.

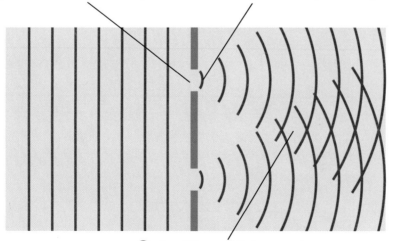
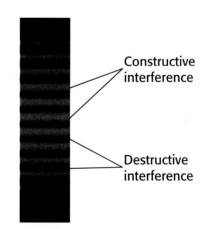

3 The diffracted light waves interfere both constructively and destructively.

4 The interference shows up on a screen as bright and dark bands.

Constructive interference

Destructive interference

Summary

- The law of reflection states that the angle of incidence is equal to the angle of reflection.

- Things that are luminous can be seen because they produce their own light. Things that are illuminated can be seen because light reflects off them.

- Absorption is the transfer of light energy to particles of matter. Scattering is an interaction of light with matter that causes light to change direction.

- Refraction of light waves can create optical illusions and can separate white light into separate colors.

- How much light waves diffract depends on the light's wavelength. Light waves diffract more when traveling through a narrow opening.

- Interference can be constructive or destructive. Interference of light waves can cause bright and dark bands.

Using Key Terms

For each pair of terms, explain how the meanings of the terms differ.

1. *refraction* and *diffraction*

2. *absorption* and *scattering*

Understanding Key Ideas

3. Which light interaction explains why you can see things that do not produce their own light?

   a. absorption   **c.** refraction

   b. reflection   **d.** scattering

4. Describe how absorption and scattering can affect a beam of light.

5. Why do objects that are underwater look closer than they actually are?

6. How does a prism separate white light into different colors?

7. What is the relationship between diffraction and the wavelength of light?

Critical Thinking

8. **Applying Concepts** Explain why you can see your reflection on a spoon but not on a piece of cloth.

9. **Making Inferences** The planet Mars does not produce light. Explain why you can see Mars shining like a star at night.

10. **Making Comparisons** Compare constructive interference and destructive interference.

Interpreting Graphics

Use the image below to answer the questions that follow.

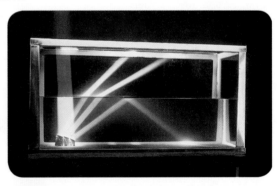

11. Why doesn't the large beam of light bend like the two beams in the middle of the tank?

12. Which light interaction explains what is happening to the bottom light beam?

For a variety of links related to this chapter, go to www.scilinks.org

Topic: Reflection and Refraction
SciLinks code: HSM1283

Light and Color

Why are strawberries red and bananas yellow? How can a soda bottle be green, yet you can still see through it?

If white light is made of all the colors of light, how do things get their color from white light? Why aren't all things white in white light? Good questions! To answer these questions, you need to know how light interacts with matter.

Light and Matter

When light strikes any form of matter, it can interact with the matter in three different ways—the light can be reflected, absorbed, or transmitted.

Reflection happens when light bounces off an object. Reflected light allows you to see things. Absorption is the transfer of light energy to matter. Absorbed light can make things feel warmer. **Transmission** is the passing of light through matter. You see the transmission of light all the time. All of the light that reaches your eyes is transmitted through air. Light can interact with matter in several ways at the same time. Look at **Figure 1.** Light is transmitted, reflected, and absorbed when it strikes the glass in a window.

What You Will Learn

- Name and describe the three ways light interacts with matter.
- Explain how the color of an object is determined.
- Explain why mixing colors of light is called *color addition*.
- Describe why mixing colors of pigments is called *color subtraction*.

Vocabulary

transmission opaque
transparent pigment
translucent

READING STRATEGY

Discussion Read this section silently. Write down questions that you have about this section. Discuss your questions in a small group.

transmission the passing of light or other form of energy through matter

Figure 1 **Transmission, Reflection, and Absorption**

You can see objects outside because light is **transmitted** through the glass.

You can see the glass and your reflection in it because light is **reflected** off the glass.

The glass feels warm when you touch it because some light is **absorbed** by the glass.

Figure 2 Transparent, Translucent, and Opaque

Transparent plastic makes it easy to see what you are having for lunch.

Translucent wax paper makes it a little harder to see exactly what's for lunch.

Opaque aluminum foil makes it impossible to see your lunch without unwrapping it.

Types of Matter

Matter through which visible light is easily transmitted is said to be **transparent.** Air, glass, and water are examples of transparent matter. You can see objects clearly when you view them through transparent matter.

Sometimes, windows in bathrooms are made of frosted glass. If you look through one of these windows, you will see only blurry shapes. You can't see clearly through a frosted window because it is translucent (trans LOO suhnt). **Translucent** matter transmits light but also scatters the light as it passes through the matter. Wax paper is an example of translucent matter.

Matter that does not transmit any light is said to be **opaque** (oh PAYK). You cannot see through opaque objects. Metal, wood, and this book are examples of opaque objects. You can compare transparent, translucent, and opaque matter in **Figure 2.**

☑ Reading Check List two examples of translucent objects. (*See the Appendix for answers to Reading Checks.*)

transparent describes matter that allows light to pass through with little interference

translucent describes matter that transmits light but that does not transmit an image

opaque describes an object that is not transparent or translucent

Colors of Objects

How is an object's color determined? Humans see different wavelengths of light as different colors. For example, humans see long wavelengths as red and short wavelengths as violet. And, some colors, like pink and brown, are seen when certain combinations of wavelengths are present.

The color that an object appears to be is determined by the wavelengths of light that reach your eyes. Light reaches your eyes after being reflected off an object or after being transmitted through an object. When your eyes receive the light, they send signals to your brain. Your brain interprets the signals as colors.

Figure 3 Opaque Objects and Color

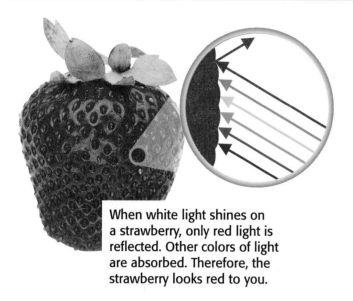

When white light shines on a strawberry, only red light is reflected. Other colors of light are absorbed. Therefore, the strawberry looks red to you.

The white hair in this cow's hide reflects all the colors of light, but the black hair absorbs all the colors.

Colors of Opaque Objects

When white light strikes a colored opaque object, some colors of light are absorbed, and some are reflected. Only the light that is reflected reaches your eyes and is detected. So, the colors of light that are reflected by an opaque object determine the color you see. For example, if a sweater reflects blue light and absorbs all other colors, you will see that the sweater is blue. Another example is shown on the left in **Figure 3.**

What colors of light are reflected by the cow shown on the right in **Figure 3**? Remember that white light includes all colors of light. So, white objects—such as the white hair in the cow's hide—appear white because all the colors of light are reflected. On the other hand, black is the absence of color. When light strikes a black object, all the colors are absorbed.

✓ Reading Check What happens when white light strikes a colored opaque object?

Colors of Transparent and Translucent Objects

The color of transparent and translucent objects is determined differently than the color of opaque objects. Ordinary window glass is colorless in white light because it transmits all the colors that strike it. But some transparent objects are colored. When you look through colored transparent or translucent objects, you see the color of light that was transmitted through the material. The other colors were absorbed, as shown in **Figure 4.**

Figure 4 *This bottle is green because the plastic transmits green light.*

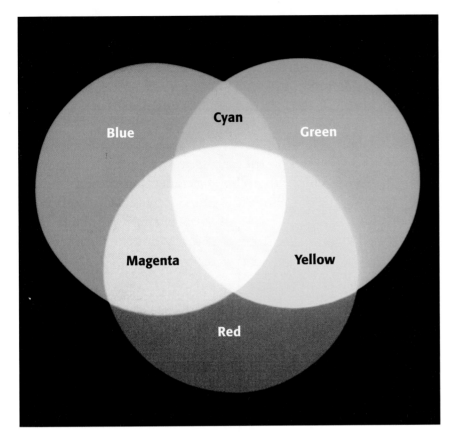

Figure 5 *Primary colors of light—written in white—combine to produce white light. Secondary colors of light—written in black—are the result of two primary colors added together.*

Mixing Colors of Light

In order to get white light, you must combine all colors of light, right? This method is one way of doing it. But you can also get light that appears white by adding just three colors of light together—red, blue, and green. The combination of these three colors is shown in **Figure 5.** In fact, these three colors can be combined in different ratios to produce many colors. Red, blue, and green are called the *primary colors of light.*

Color Addition

When colors of light combine, you see different colors. Combining colors of light is called *color addition.* When two primary colors of light are added together, you see a *secondary color of light.* The secondary colors of light are cyan (blue plus green), magenta (blue plus red), and yellow (red plus green). **Figure 5** shows how secondary colors of light are formed.

Light and Color Television

The colors on a color television are produced by color addition of the primary colors of light. A television screen is made up of groups of tiny red, green, and blue dots. Each dot will glow when the dot is hit by an electron beam. The colors given off by the glowing dots add together to produce all the different colors you see on the screen.

Television Colors

Turn on a color television. Ask an adult to carefully sprinkle a few tiny drops of water onto the television screen. Look closely at the drops of water, and discuss what you see. In your **science journal**, write a description of what you saw.

Mixing Colors of Pigment

If you have ever tried mixing paints in art class, you know that you can't make white paint by mixing red, blue, and green paint. The difference between mixing paint and mixing light is due to the fact that paint contains pigments.

Pigments and Color

A **pigment** is a material that gives a substance its color by absorbing some colors of light and reflecting others. Almost everything contains pigments. Chlorophyll (KLAWR uh FIL) and melanin (MEL uh nin) are two examples of pigments. Chlorophyll gives plants a green color, and melanin gives your skin its color.

✓ **Reading Check** What is a pigment?

Color Subtraction

Each pigment absorbs at least one color of light. Look at **Figure 6.** When you mix pigments together, more colors of light are absorbed or taken away. So, mixing pigments is called *color subtraction.*

The *primary pigments* are yellow, cyan, and magenta. They can be combined to produce any other color. In fact, every color in this book was produced by using just the primary pigments and black ink. The black ink was used to provide contrast to the images. **Figure 7** shows how the four pigments combine to produce many different colors.

Figure 6 *Primary pigments—written in black—combine to produce black. Secondary pigments—written in white—are the result of the subtraction of two primary pigments.*

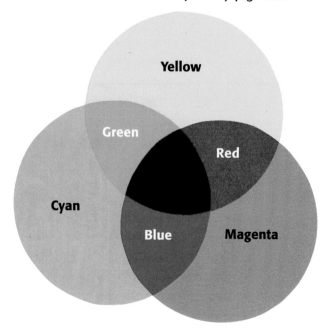

Quick Lab

Rose-Colored Glasses?

1. Obtain **four plastic filters**—red, blue, yellow, and green.

2. Look through one filter at an object across the room. Describe the object's color.

3. Repeat step 2 with each of the filters.

4. Repeat step 2 with two or three filters together.

5. Why do you think the colors change when you use more than one filter?

6. Write your observations and answers.

Figure 7 Color Subtraction and Color Printing

The picture of the balloon on the left was made by overlapping yellow ink, cyan ink, magenta ink, and black ink.

Yellow   Cyan   Magenta   Black

SECTION Review

Summary

- Objects are transparent, translucent, or opaque, depending on their ability to transmit light.

- Colors of opaque objects are determined by the color of light that they reflect.

- Colors of translucent and transparent objects are determined by the color of light they transmit.

- White light is a mixture of all colors of light.

- Light combines by color addition. The primary colors of light are red, blue, and green.

- Pigments give objects color. Pigments combine by color subtraction. The primary pigments are magenta, cyan, and yellow.

Using Key Terms

1. Use the following terms in the same sentence: *transmission* and *transparent*.

2. In your own words, write a definition for each of the following terms: *translucent* and *opaque*.

Understanding Key Ideas

3. You can see through a car window because the window is

   a. opaque.   **c.** transparent.
   b. translucent.   **d.** transmitted.

4. Name and describe three different ways light interacts with matter.

5. How is the color of an opaque object determined?

6. Describe how the color of a transparent object is determined.

7. What are the primary colors of light, and why are they called *primary colors*?

8. What four colors of ink were used to print this book?

Critical Thinking

9. **Applying Concepts** What happens to the different colors of light when white light shines on an opaque violet object?

10. **Analyzing Ideas** Explain why mixing colors of light is called *color addition* but mixing pigments is called *color subtraction*.

Interpreting Graphics

11. Look at the image below. The red rose was photographed in red light. Explain why the leaves appear black and the petals appear red.

For a variety of links related to this chapter, go to www.scilinks.org

Topic: Colors
SciLinks code: HSM0314

Skills Practice Lab

Use flashlights to mix colors of light by color addition.

Use paints to mix colors of pigments by color subtraction.

Part A
• colored filters, red, green, and blue (1 of each)
• flashlights (3)
• paper, white
• tape, masking

Part B
• cups, small plastic or paper (2)
• paintbrush
• paper, white
• ruler, metric
• tape, masking
• water
• watercolor paints

Mixing Colors

Mix two colors, such as red and green, and you create a new color. Is the new color brighter or darker? Color and brightness depend on the light that reaches your eye. And what reaches your eye depends on whether you are adding colors (mixing colors of light) or subtracting colors (mixing colors of pigments). In this activity, you will do both types of color formation and see the results firsthand!

Part A: Color Addition

Procedure

1 Tape a colored filter over each flashlight lens.

2 In a darkened room, shine the red light on a sheet of white paper. Then, shine the green light next to the red light. You should have two circles of light, one red and one green, next to each other.

3 Move the flashlights so that the circles overlap by half their diameter. What color is formed where the circles overlap? Is the mixed area brighter or darker than the single-color areas? Record your observations.

Red ? Green

4 Repeat steps 2 and 3 with the red and blue lights.

5 Now, shine all three lights at the same point on the paper. Record your observations.

Analyze the Results

1 Describing Events In general, when you mixed two colors, was the result brighter or darker than the original colors?

2 Explaining Events In step 5, you mixed all three colors. Was the resulting color brighter or darker than when you mixed two colors? Explain your observations in terms of color addition.

Draw Conclusions

3 Making Predictions What do you think would happen if you mixed together all the colors of light? Explain your answer.

Part B: Color Subtraction

Procedure

1 Place a piece of masking tape on each cup. Label one cup "Clean" and the other cup "Dirty." Fill each cup about half full with water.

2 Wet the paintbrush thoroughly in the "Clean" cup. Using the watercolor paints, paint a red circle on the white paper. The circle should be approximately 4 cm in diameter.

3 Clean the brush by rinsing it first in the "Dirty" cup and then in the "Clean" cup.

4 Paint a blue circle next to the red circle. Then, paint half the red circle with the blue paint.

5 Examine the three areas: red, blue, and mixed. What color is the mixed area? Does it appear brighter or darker than the red and blue areas? Record your observations.

6 Clean the brush by repeating Step 3. Paint a green circle 4 cm in diameter, and then paint half the blue circle with green paint.

7 Examine the green, blue, and mixed areas. Record your observations.

8 Now add green paint to the mixed red-blue area so that you have an area that is a mixture of red, green, and blue paint. Clean the brush again.

9 Finally, record your observations of this new mixed area.

Analyze the Results

1 Identifying Patterns In general, when you mixed two colors, was the result brighter or darker than the original colors?

2 Analyzing Results In step 8, you mixed all three colors. Was the result brighter or darker than the result from mixing two colors? Explain what you saw in terms of color subtraction.

Draw Conclusions

3 Drawing Conclusions Based on your results, what do you think would happen if you mixed all the colors of paint? Explain your answer.

Chapter Review

USING KEY TERMS

Complete each of the following sentences by choosing the correct term from the word bank.

interference	radiation
scattering	opaque
translucent	transmission
electromagnetic wave	electromagnetic spectrum

1 _____ is the transfer of energy by electromagnetic waves.

2 This book is a(n) _____ object.

3 _____ is a wave interaction that occurs when two or more waves overlap and combine.

4 Light is a kind of _____ and can therefore travel through matter and space.

5 During _____, light travels through an object.

UNDERSTANDING KEY IDEAS

Multiple Choice

6 Electromagnetic waves transmit
  a. charges.
  b. fields.
  c. matter.
  d. energy.

7 Objects that transmit light easily are
  a. opaque.
  b. translucent.
  c. transparent.
  d. colored.

8 You can see yourself in a mirror because of
  a. absorption.
  b. scattering.
  c. regular reflection.
  d. diffuse reflection.

9 Shadows have blurry edges because of
  a. diffraction.
  b. scattering.
  c. diffuse reflection.
  d. refraction.

10 What color of light is produced when red light is added to green light?
  a. cyan   c. yellow
  b. blue   d. white

11 Prisms produce the colors of the rainbow through
  a. reflection.   c. diffraction.
  b. refraction.   d. interference.

12 Which kind of electromagnetic wave travels fastest in a vacuum?
  a. radio wave
  b. visible light
  c. gamma ray
  d. They all travel at the same speed.

13 Electromagnetic waves are made of
  a. vibrating particles.
  b. vibrating charged particles.
  c. vibrating electric and magnetic fields.
  d. All of the above

Short Answer

14 How are gamma rays used?

15 What are two uses for radio waves?

16 Why is it difficult to see through glass that has frost on it?

Math Skills

17 Calculate the time it takes for light from the sun to reach Mercury. Mercury is 54,900,000 km away from the sun.

CRITICAL THINKING

18 **Concept Mapping** Use the following terms to create a concept map: *light, matter, reflection, absorption,* and *transmission.*

19 **Applying Concepts** A tern is a type of bird that dives underwater to catch fish. When a young tern begins learning to catch fish, the bird is rarely successful. The tern has to learn that when a fish appears to be in a certain place underwater, the fish is actually in a slightly different place. Why does the tern see the fish in the wrong place?

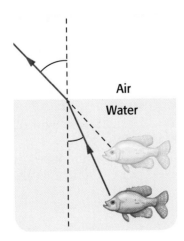

Air

Water

20 **Evaluating Conclusions** Imagine that you are teaching your younger brother about light. You tell him that white light is light of all the colors of the rainbow combined. But your brother says that you are wrong because mixing different colors of paint produces black and not white. Explain why your brother's conclusion is wrong.

21 **Making Inferences** If you look around a parking lot during the summer, you might see sunshades set up in the windshields of cars. How do sunshades help keep the insides of cars cool?

INTERPRETING GRAPHICS

22 Each of the pictures below shows the effects of a wave interaction of light. Identify the interaction involved.

a.

b.

c.

Standardized Test Preparation

Read each of the passages below. Then, answer the questions that follow each passage.

Passage 1 Jaundice occurs in some infants when bilirubin—a pigment in healthy red blood cells—builds up in the bloodstream as blood cells break down. This excess bilirubin is deposited in the skin, giving the skin a yellowish hue. Jaundice is not dangerous if treated quickly. If left untreated, it can lead to brain damage.

The excess bilirubin in the skin is best broken down by bright blue light. For this reason, hospitals hang special blue fluorescent lights above the cribs of newborns needing treatment. The blue light is sometimes balanced with light of other colors so that doctors and nurses can be sure the baby is not blue from a lack of oxygen.

1. Which of the following is a fact in the passage?

A Jaundice is always very dangerous.

B Bilirubin in the skin of infants can be broken down with bright blue light.

C Excess bilirubin in the skin gives the skin a bright blue hue.

D Blue lights can make a baby blue from a lack of oxygen.

2. What is the purpose of this passage?

F to explain what jaundice is and how it is treated

G to warn parents about shining blue light on their babies

H to persuade light bulb manufacturers to make blue light bulbs

I to explain the purpose of bilirubin in red blood cells

Passage 2 If you have ever looked inside a toaster while toasting a piece of bread, you may have seen thin wires or bars glowing red. The wires give off energy as light when heated to a high temperature. Light produced by hot objects is called *incandescent light*. Most of the lamps in your home probably use incandescent light bulbs.

Sources of incandescent light also release a large amount of <u>thermal</u> energy. Thermal energy is sometimes called *heat energy*. Sometimes, thermal energy from incandescent light is used to cook food or to warm a room. But often this thermal energy is not used for anything. For example, the thermal energy given off by light bulbs is not very useful.

1. What does the word *thermal* mean, based on its use in the passage?

A light

B energy

C heat

D food

2. What is incandescent light?

F light used for cooking food

G light that is red in color

H light that is not very useful

I light produced by hot objects

3. Which of the following can be inferred from the passage?

A Sources of incandescent light are rarely found in an average home.

B A toaster uses thermal energy to toast bread.

C Incandescent light from light bulbs is often used to cook food.

D The thermal energy produced by incandescent light sources is always useful.

The angles of refraction in the table were measured when a beam of light entered the material from air at a 45° angle. Use the table below to answer the questions that follow.

Material and Refraction		
Material	Index of refraction	Angle of refraction
Diamond	2.42	17°
Glass	1.52	28°
Quartz	1.46	29°
Water	1.33	32°

1. Which material has the highest index of refraction?

   A diamond
   B glass
   C quartz
   D water

2. Which material has the greatest angle of refraction?

   F diamond
   G glass
   H quartz
   I water

3. Which of the following statements **best** describes the data in the table?

   A The higher the index of refraction, the greater the angle of refraction.
   B The higher the index of refraction, the smaller the angle of refraction.
   C The greater the angle of refraction, the higher the index of refraction.
   D There is no relationship between the index of refraction and the angle of refraction.

4. Which two materials would be the most difficult to separate by observing only their angles of refraction?

   F diamond and glass
   G glass and quartz
   H quartz and water
   I water and diamond

Read each question below, and choose the best answer.

1. A square metal plate has an area of 46.3 cm². The length of one side of the plate is between which two values?

   A 4 cm and 5 cm
   B 5 cm and 6 cm
   C 6 cm and 7 cm
   D 7 cm and 8 cm

2. A jet was flying over the Gulf of Mexico at an altitude of 2,150 m. Directly below the jet, a submarine was at a depth of −383 m. What was the distance between the jet and the submarine?

   F −2,533 m
   G −1,767 m
   H 1,767 m
   I 2,533 m

3. The speed of light in a vacuum is exactly 299,792,458 m/s. Which of the following is a good estimate of the speed of light?

   A 3.0×10^{-8} m/s
   B 2.0×10^{8} m/s
   C 3.0×10^{8} m/s
   D 3.0×10^{9} m/s

4. The wavelength of the yellow light produced by a sodium vapor lamp is 0.000000589 m. Which of the following is equal to the wavelength of the sodium lamp's yellow light?

   F -5.89×10^{7} m
   G 5.89×10^{-9} m
   H 5.89×10^{-7} m
   I 5.89×10^{7} m

5. Amira purchased a box of light bulbs for $3.81. There are three light bulbs in the box. What is the cost per light bulb?

   A $0.79
   B $1.06
   C $1.27
   D $11.43

Standardized Test Preparation

Science in Action

Weird Science

Fireflies Light the Way

Just as beams of light from lighthouses warn boats of approaching danger, the light of an unlikely source—fireflies—is being used by scientists to warn food inspectors of bacterial contamination.

Fireflies use an enzyme called *luciferase* to make light. Scientists have taken the gene from fireflies that tells cells how to make luciferase. They put this gene into a virus that preys on bacteria. The virus is not harmful to humans and can be mixed into meat. When the virus infects bacteria in the meat, the virus transfers the gene into the genes of the bacteria. The bacteria then produce luciferase and glow! So, if a food inspector sees glowing meat, the inspector knows that the meat is contaminated with bacteria.

Science, Technology, and Society

It's a Heat Wave

In 1946, Percy Spencer visited a laboratory belonging to Raytheon—the company he worked for. When he stood near a device called a *magnetron,* he noticed that a candy bar in his pocket melted. Spencer hypothesized that the microwaves produced by the magnetron caused the candy bar to warm up and melt. To test his hypothesis, Spencer put a bag of popcorn kernels next to the magnetron. The microwaves heated the kernels, causing them to pop! Spencer's simple experiment showed that microwaves could heat foods quickly. Spencer's discovery eventually led to the development of the microwave oven—an appliance found in many kitchens today.

Social Studies  ACTIVITY

WRITING SKILL Many cultures have myths to explain certain natural phenomena. Read some of these myths. Then, write your own myth titled "How Fireflies Got Their Fire."

Math ACTIVITY

Popcorn pops when the inside of the kernel reaches a temperature of about 175°C. Convert this temperature to degrees Fahrenheit.

Albert Einstein

A Light Pioneer When Albert Einstein was 15 years old, he asked himself, "What would the world look like if I were speeding along on a motorcycle at the speed of light?" For many years afterward, he would think about this question and about the very nature of light, time, space, and matter. He even questioned the ideas of Isaac Newton, which had been widely accepted for 200 years. Einstein was bold. And he was able to see the universe in a totally new way.

In 1905, Einstein published a paper on the nature of light. He knew from the earlier experiments of others that light was a wavelike phenomenon. But he theorized that light could also travel as particles. Scientists did not readily accept Einstein's particle theory of light. Even 10 years later, the American physicist Robert Millikan, who proved that the particle theory of light was true, was reluctant to believe his own experimental results. Einstein's theory helped pave the way for television, computers, and other important technologies. The theory also earned Einstein a Nobel Prize in physics in 1921.

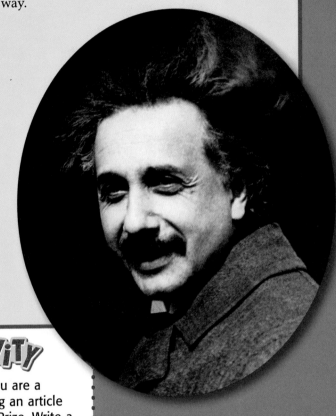

Language Arts ACTiViTY

WRITING SKILL Imagine that it is 1921. You are a newspaper reporter writing an article about Albert Einstein and his Nobel Prize. Write a one-page article about Albert Einstein, his theory, and the award he won.

go.hrw.com

To learn more about these Science in Action topics, visit **go.hrw.com** and type in the keyword **HP5LGTF**.

Current Science

Check out Current Science® articles related to this chapter by visiting go.hrw.com. Just type in the keyword HP5CS22.

Light and Our World

The Big Idea

Mirrors and lenses change the path of light waves and affect the images that you see.

About the PHOTO

This photo of Earth was taken by a satellite in space. All of the dots of light in this photo are lights in cities around the world. In areas with many dots, people live in cities that are close together. Light is very important in your everyday life. Not only does light help you see at night but light waves can also be used to send information over long distances. In fact, the satellite that took this picture sent the picture to Earth by using light waves!

PRE-READING ACTIVITY

FOLDNOTES **Tri-Fold** Before you read the chapter, create the FoldNote entitled "Tri-Fold" described in the **Study Skills** section of the Appendix. Write what you know about light in the column labeled "Know." Then, write what you want to know in the column labeled "Want." As you read the chapter, write what you learn about light in the column labeled "Learn."

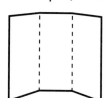

START-UP ACTIVITY

Mirror, Mirror

In this activity, you will study images formed by flat, or plane, mirrors.

Procedure

1. Tape a sheet of **graph paper** on your desk. Stand a **flat mirror** in the middle of the paper. Hold the mirror in place with pieces of **modeling clay.**

2. Place a **pen** four squares in front of the mirror. How many squares behind the mirror is the image of the pen? Move the pen farther away from the mirror. How did the image change?

3. Replace the mirror with **colored glass.** Look at the image of the pen in the glass. Compare the image in the glass with the one in the mirror.

4. Draw a square on the graph paper in front of the glass. Then, look through the glass, and trace the image of the square on the paper behind the glass. Using a **metric ruler,** measure and compare the two squares.

Analysis

1. How does the distance from an object to a plane mirror compare with the apparent distance from the mirror to the object's image behind the mirror?

2. Images formed in the colored glass are similar to images formed in a plane mirror. In general, how does the size of an object compare with that of its image in a plane mirror?

Mirrors and Lenses

When walking by an ambulance, you notice that the letters on the front of the ambulance look strange. Some letters are backward, and they don't seem to spell a word!

Look at **Figure 1.** The letters spell the word *ambulance* when viewed in a mirror. Images in mirrors are reversed left to right. The word *ambulance* is spelled backward so that people driving cars can read it when they see an ambulance in their rearview mirrors. To understand how images are formed in mirrors, you must first learn how to use rays to trace the path of light waves.

Rays and the Path of Light Waves

Light waves are electromagnetic waves. Light waves travel from their source in all directions. If you could trace the path of one light wave as it travels away from a light source, you would find that the path is a straight line. Because light waves travel in straight lines, you can use an arrow called a *ray* to show the path and the direction of a light wave.

Rays and Reflected and Refracted Light

Rays help to show the path of a light wave after it bounces or bends. Light waves that bounce off an object are reflected. Light waves that bend when passing from one medium to another are refracted. So, rays in ray diagrams show changes in the direction light travels after being reflected by mirrors or refracted by lenses.

What You Will Learn

- Use ray diagrams to show how light is reflected or refracted.
- Compare plane mirrors, concave mirrors, and convex mirrors.
- Use ray diagrams to show how mirrors form images.
- Describe the images formed by concave and convex lenses.

Vocabulary

plane mirror	lens
concave mirror	convex lens
convex mirror	concave lens

READING STRATEGY

Reading Organizer As you read this section, make a concept map by using the terms above.

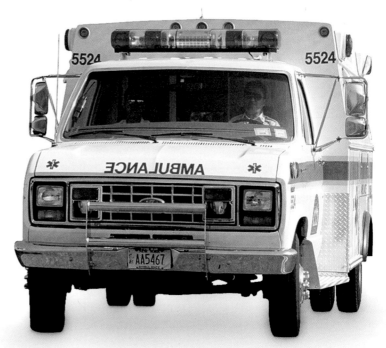

Figure 1 *If you hold this photo up to the mirror in your bathroom, you will see the word* AMBULANCE.

Mirrors and Reflection of Light

Have you ever looked at your reflection in a metal spoon? The spoon is like a mirror but not like a bathroom mirror! If you look on one side of the spoon, your face is upside down. But on the other side, your face is right side up. Why? Read on to find out!

The shape of a mirror affects the way light reflects from it. So, the image you see in your bathroom mirror differs from the image you see in a spoon. Mirrors are classified by their shape. Three shapes of mirrors are plane, concave, and convex.

Plane Mirrors

Most mirrors, such as the one in your bathroom, are plane mirrors. A **plane mirror** is a mirror that has a flat surface. When you look in a plane mirror, your reflection is right side up. The image is also the same size as you are. Images in plane mirrors are reversed left to right, as shown in **Figure 2.**

In a plane mirror, your image appears to be the same distance behind the mirror as you are in front of it. Why does your image seem to be behind the mirror? When light reflects off the mirror, your brain thinks the reflected light travels in a straight line from behind the mirror. The ray diagram in **Figure 3** explains how light travels when you look into a mirror. The image formed by a plane mirror is a virtual image. A *virtual image* is an image through which light does not travel.

✓ Reading Check What is a virtual image? (*See the Appendix for answers to Reading Checks.*)

Figure 2 *Rearview mirrors in cars are plane mirrors. This mirror shows the reflection of the front of the ambulance shown in Figure 1.*

plane mirror a mirror that has a flat surface

Figure 3 **How Images Are Formed in Plane Mirrors**

The rays show how light reaches your eyes. The dotted lines show where the light appears to come from.

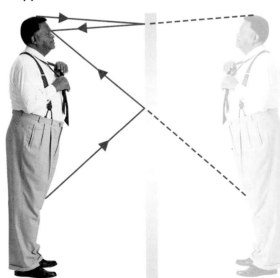

Light reflects off you and strikes the mirror. The light then reflects off the mirror at an angle equal to the angle at which the light hit the mirror. Some of the reflected light enters your eyes.

Your image appears to be behind the mirror because your brain assumes that the light rays that enter your eyes travel in a straight line from an object to your eyes.

Figure 4 *Concave mirrors are curved like the inside of a spoon. The image formed by a concave mirror depends on the optical axis, focal point, and focal length of the mirror.*

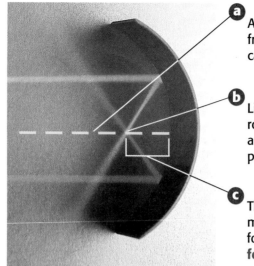

a A straight line drawn outward from the center of the mirror is called the **optical axis**.

b Light rays entering the mirror parallel to the optical axis are reflected through a single point, called the **focal point**.

c The distance between the mirror's surface and the focal point is called the **focal length**.

concave mirror a mirror that is curved inward like the inside of a spoon

convex mirror a mirror that is curved outward like the back of a spoon

Concave Mirrors

A mirror that is curved inward is called a **concave mirror.** The images formed by concave mirrors differ from the images formed by plane mirrors. The image formed by a concave mirror depends on three things: the optical axis, focal point, and focal length of the mirror. **Figure 4** explains these terms.

You have already learned that plane mirrors can form only virtual images. Concave mirrors also form virtual images. But they can form real images, too. A *real image* is an image through which light passes. A real image can be projected onto a screen, but a virtual image cannot.

Concave Mirrors and Ray Diagrams

To find out what kind of image a concave mirror forms, you can make a ray diagram. Draw two rays from the top of the object to the mirror. Then, draw rays reflecting from the surface of the mirror. If the reflected rays cross in front of the mirror, a real image is formed. If the reflected rays do not cross in front of the mirror, extend the reflected rays in straight lines behind the mirror. Those lines will cross to show where a virtual image is formed. Study **Figure 5** to better understand ray diagrams.

If an object is placed at the focal point of a concave mirror, no image will form. All rays that pass through the focal point on their way to the mirror will reflect parallel to the optical axis. The rays will never cross in front of or behind the mirror. If you put a light source at the focal point of a concave mirror, light will reflect outward in a powerful beam. So, concave mirrors are used in car headlights and flashlights.

For another activity related to this chapter, go to **go.hrw.com** and type in the keyword **HP5LOWW.**

Reading Check How can a concave mirror be used to make a powerful beam of light?

Figure 5 *The type of image formed by a concave mirror depends on the distance between the object and the mirror.*

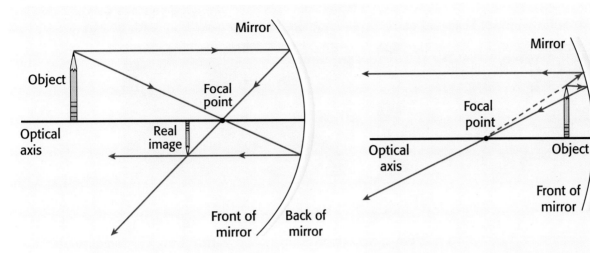

An object more than 1 focal length away from a concave mirror forms an **upside-down, real image.**

An object less than 1 focal length away from a concave mirror forms a **right-side-up, virtual image.** The dotted lines trace the reflected rays behind the mirror to find the virtual image.

Convex Mirrors

If you look at your reflection in the back of a spoon, you will notice that your image is right side up and small. The back of a spoon is a convex mirror. A **convex mirror** is a mirror that curves outward. **Figure 6** shows how an image is formed by a convex mirror. The reflected rays do not cross in front of a convex mirror. So, the reflected rays are extended behind the mirror to find the virtual image. All images formed by convex mirrors are virtual, right side up, and smaller than the original object. Convex mirrors are useful because they make images of large areas. So, convex mirrors are often used for security in stores and factories. Convex mirrors are also used as side mirrors on cars and trucks.

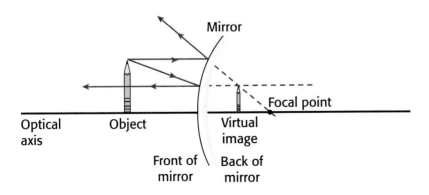

Figure 6 *All images formed by convex mirrors are formed behind the mirror. Therefore, all images formed by convex mirrors are virtual.*

SCHOOL to HOME

Car Mirrors

Sit in the passenger side of a car. Ask an adult at home to stand one car-length behind the car. Look at the adult's reflection in the passenger side mirror. Then, look at the adult's reflection in the rearview mirror. Make a table comparing the two mirrors and the images you saw in each mirror.

ACTIVITY

Figure 7 **How Light Passes Through Lenses**

When light rays pass through a **convex lens,** the rays are refracted toward each other.

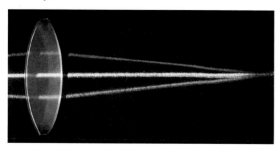

When light rays pass through a **concave lens,** the rays are refracted away from each other.

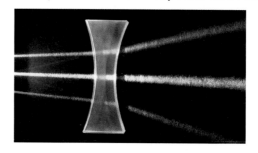

Lenses and Refraction of Light

What do cameras, telescopes, and movie projectors have in common? They all use lenses to create images. A **lens** is a transparent object that forms an image by refracting, or bending, light. Lenses are classified by their shape. Two kinds of lenses, convex and concave, are shown in **Figure 7.** The yellow beams in **Figure 7** show that light rays that pass through the center of any lens are not refracted. Like mirrors, lenses have a focal point and an optical axis.

lens a transparent object that refracts light waves such that they converge or diverge to create an image

convex lens a lens that is thicker in the middle than at the edges

concave lens a lens that is thinner in the middle than at the edges

Convex Lenses

A **convex lens** is a lens that is thicker in the middle than at the edges. Convex lenses form different kinds of images. The ways in which two of these kinds of images are formed are shown in **Figure 8.** In addition, a convex lens can form a real image that is larger than the object if the object is between 1 and 2 focal lengths away from the lens. Convex lenses have many uses. For example, magnifying lenses and camera lenses are convex lenses. And convex lenses are sometimes used in eyeglasses.

Figure 8 *The distance between an object and a convex lens determines the size and the kind of image formed.*

✓ **Reading Check** What is a convex lens?

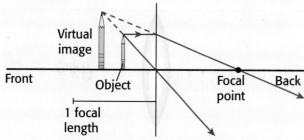

If an object is less than 1 focal length away from a convex lens, a **virtual image** is formed. The image is larger than the object.

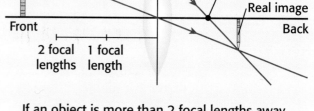

If an object is more than 2 focal lengths away from the lens, a **real image** is formed. The image is smaller than the object.

Concave Lenses

A **concave lens** is a lens that is thinner in the middle than at the edges. Light rays entering a concave lens parallel to the optical axis always bend away from each other and appear to come from a focal point in front of the lens. The rays never meet. So, concave lenses never form a real image. Instead, they form virtual images, as shown in **Figure 9.** Concave lenses are sometimes combined with other lenses in telescopes. The combination of lenses produces clearer images of distant objects. Concave lenses are also used in microscopes and eyeglasses.

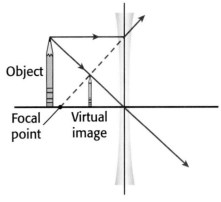

Figure 9 *Concave lenses form virtual images. The image is smaller than the object.*

SECTION Review

Summary

- Rays are arrows that show the path of a single light wave.
- Ray diagrams can be used to find where images are formed by mirrors and lenses.
- Plane mirrors and convex mirrors produce virtual images. Concave mirrors produce both real images and virtual images.
- Convex lenses produce both real images and virtual images. Concave lenses produce only virtual images.

Using Key Terms

For each pair of terms, explain how the meanings of the terms differ.

1. *convex mirror* and *concave mirror*

2. *convex lens* and *concave lens*

Understanding Key Ideas

3. Which of the following can form real images?
   a. a plane mirror
   b. a convex mirror
   c. a convex lens
   d. a concave lens

4. Explain how you can use a ray diagram to determine if a real image or a virtual image is formed by a mirror.

5. Compare the images formed by plane mirrors, concave mirrors, and convex mirrors.

6. Describe the images that can be formed by convex lenses.

7. Explain why a concave lens cannot form a real image.

Critical Thinking

8. **Applying Concepts** Why is an image right side up on the back of a spoon but upside down on the inside of a spoon?

9. **Making Inferences** Teachers sometimes use overhead projectors to show transparencies on a screen. What type of lens does an overhead projector use?

Interpreting Graphics

10. Look at the ray diagram below. Identify the type of lens and the kind of image that is formed.

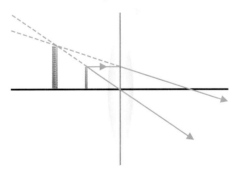

For a variety of links related to this chapter, go to www.scilinks.org

Topic: Mirrors; Lenses
SciLinks code: HSM0970; HSM0868

Light and Sight

When you look around, you can see objects both near and far. You can also see the different colors of the objects.

You see objects that produce their own light because the light is detected by your eyes. You see all other objects because light reflected from the objects enters your eyes. But how do your eyes work, and what causes people to have vision problems?

How You Detect Light

Visible light is the part of the electromagnetic spectrum that can be detected by your eyes. Your eye gathers light to form the images that you see. The steps of this process are shown in **Figure 1.** Muscles around the lens change the thickness of the lens so that objects at different distances can be seen in focus. The light that forms the real image is detected by receptors in the retina called *rods* and *cones*. Rods can detect very dim light. Cones detect colors in bright light.

What You Will Learn

● Identify the parts of the human eye, and describe their functions.
● Describe three common vision problems.
● Describe surgical eye correction.

Vocabulary
nearsightedness
farsightedness

READING STRATEGY

Reading Organizer As you read this section, make a flowchart of how the eye works.

Figure 1 How Your Eyes Work

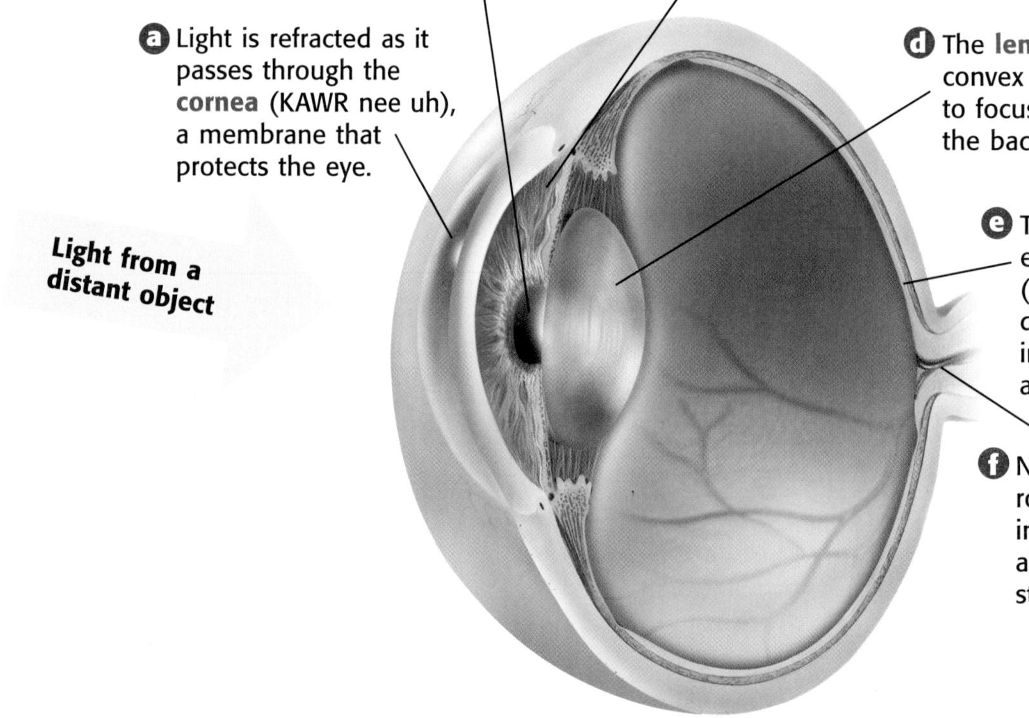

ⓑ Light passes through the **pupil,** the opening in the eye.

ⓒ The size of the pupil is controlled by the **iris,** which is the colored part of the eye.

ⓐ Light is refracted as it passes through the **cornea** (KAWR nee uh), a membrane that protects the eye.

ⓓ The **lens** of the eye is convex and refracts light to focus a real image on the back of the eye.

ⓔ The back surface of the eye is called the **retina** (RET 'n uh). Light is detected by receptors in the retina called *rods* and *cones.*

Light from a distant object

ⓕ Nerves attached to the rods and cones carry information to the brain about the light that strikes the retina.

Figure 2 **Correcting Nearsightedness and Farsightedness**

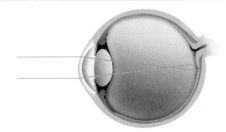

Nearsightedness happens when the eye is too long, which causes the lens to focus light in front of the retina.

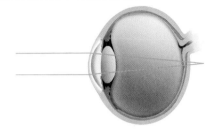

Farsightedness happens when the eye is too short, which causes the lens to focus light behind the retina.

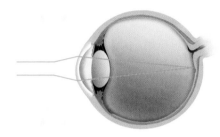

A **concave lens** placed in front of a nearsighted eye refracts the light outward. The lens in the eye can then focus the light on the retina.

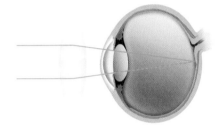

A **convex lens** placed in front of a farsighted eye focuses the light. The lens in the eye can then focus the light on the retina.

Common Vision Problems

People who have normal vision can clearly see objects that are close and objects that are far away. They can also tell the difference between all colors of visible light. But because the eye is complex, it's no surprise that many people have defects in their eyes that affect their vision.

Nearsightedness and Farsightedness

The lens of a properly working eye focuses light on the retina. So, the images formed are always clear. Two common vision problems happen when light is not focused on the retina, as shown in **Figure 2. Nearsightedness** happens when a person's eye is too long. A nearsighted person can see something clearly only if it is nearby. Objects that are far away look blurry. **Farsightedness** happens when a person's eye is too short. A farsighted person can see faraway objects clearly. But things that are nearby look blurry. **Figure 2** also shows how these vision problems can be corrected with glasses.

✓ Reading Check What causes nearsightedness and farsightedness? (*See the Appendix for answers to Reading Checks.*)

nearsightedness a condition in which the lens of the eye focuses distant objects in front of rather than on the retina

farsightedness a condition in which the lens of the eye focuses distant objects behind rather than on the retina

Figure 3 *The photo on the left is what a person who has normal vision sees. The photo on the right is a simulation of what a person who has red-green color deficiency might see.*

Color Deficiency

About 5% to 8% of men and 0.5% of women in the world have *color deficiency,* or colorblindness. The majority of people who have color deficiency can't tell the difference between shades of red and green or can't tell red from green. **Figure 3** compares what a person with normal vision sees with what a person who has red-green color deficiency sees. Color deficiency cannot be corrected.

Color deficiency happens when the cones in the retina do not work properly. The three kinds of cones are named for the colors they detect most—red, green, or blue. But each kind can detect many colors of light. A person who has normal vision can see all colors of visible light. But in some people, the cones respond to the wrong colors. Those people see certain colors, such as red and green, as a different color, such as yellow.

✔ Reading Check What are the three kinds of cones?

CONNECTION TO Biology

Color Deficiency and Genes The ability to see color is a sex-linked genetic trait. Certain genes control which colors of light the cones detect. If these genes are defective in a person, that person will have color deficiency. A person needs one set of normal genes to have normal color vision. Genes that control the red cones and the green cones are on the X chromosome. Women have two X chromosomes, but men have only one. So, men are more likely than women to lack a set of these genes and to have red-green color deficiency. Research two other sex-linked traits, and make a graph comparing the percentage of men and women who have the traits.

ACTIVITY

Surgical Eye Correction

Using surgery to correct nearsightedness or farsightedness is possible. Surgical eye correction works by reshaping the patient's cornea. Remember that the cornea refracts light. So, reshaping the cornea changes how light is focused on the retina.

To prepare for eye surgery, an eye doctor uses a machine to measure the patient's corneas. A laser is then used to reshape each cornea so that the patient gains perfect or nearly perfect vision. **Figure 4** shows a patient undergoing eye surgery.

Risks of Surgical Eye Correction

Although vision-correction surgery can be helpful, it has some risks. Some patients report glares or double vision. Others have trouble seeing at night. Other patients lose vision permanently. People under 20 years old shouldn't have vision-correction surgery because their vision is still changing.

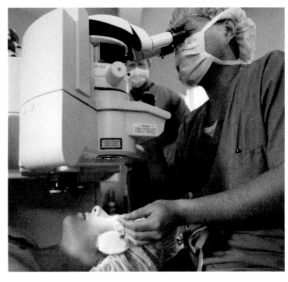

Figure 4 *An eye surgeon uses a very precise laser to reshape this patient's cornea.*

SECTION Review

Summary

- The human eye has several parts, including the cornea, the pupil, the iris, the lens, and the retina.

- Nearsightedness and farsightedness happen when light is not focused on the retina. Both problems can be corrected with glasses or eye surgery.

- Color deficiency is a condition in which cones in the retina respond to the wrong colors.

- Eye surgery can correct some vision problems.

Using Key Terms

1. Use each of the following terms in a separate sentence: *nearsightedness* and *farsightedness*.

Understanding Key Ideas

2. A person who is nearsighted will have the most trouble reading
   a. a computer screen in front of him or her.
   b. a book in his or her hands.
   c. a street sign across the street.
   d. the title of a pamphlet on a nearby table.

3. List the parts of the eye, and describe what each part does.

4. What are three common vision problems?

5. How are nearsightedness and farsightedness corrected?

6. Describe surgical eye correction.

7. What do the rods and cones in the eye do?

Math Skills

8. About 0.5% of women have a color deficiency. How many women out of 200 have a color deficiency?

Critical Thinking

9. **Forming Hypotheses** Why do you think color deficiency cannot be corrected?

10. **Expressing Opinions** Would you have surgical eye correction? Explain your reasons.

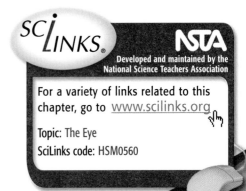

SCI LINKS®

Developed and maintained by the National Science Teachers Association

For a variety of links related to this chapter, go to www.scilinks.org

Topic: The Eye
SciLinks code: HSM0560

Light and Technology

What do cameras, telescopes, lasers, cellular telephones, and satellite televisions have in common?

They are all types of technology that use light or other electromagnetic waves. Read on to learn how these and other types of light technology are useful in your everyday life.

Optical Instruments

Optical instruments are devices that use mirrors and lenses to help people make observations. Some optical instruments help you see things that are very far away. Others help you see things that are very small. Some optical instruments record images. The optical instrument that you are probably most familiar with is the camera.

Cameras

Cameras are used to record images. **Figure 1** shows the parts of a 35 mm camera. A digital camera has a lens, a shutter, and an aperture (AP uhr chuhr) like a 35 mm camera has. But instead of using film, a digital camera uses light sensors to record images. The sensors send an electrical signal to a computer in the camera. This signal contains data about the image that is stored in the computer, on a memory stick, card, or disk.

What You Will Learn

- Describe three optical instruments.
- Explain what laser light is, and identify uses for lasers.
- Describe how optical fibers work.
- Explain polarized light.
- Explain how radio waves and microwaves are used in four types of communication technology.

Vocabulary

laser
hologram

READING STRATEGY

Prediction Guide Before reading this section, write the title of each heading in this section. Next, under each heading, write what you think you will learn.

Figure 1   How a Camera Works

The **shutter** opens and closes behind the lens to control how much light enters the camera. The longer the shutter is open, the more light enters the camera.

The **lens** of a camera is a convex lens that focuses light on the film. Moving the lens focuses light from objects at different distances.

The **film** is coated with chemicals that react when they are exposed to light. The result is an image stored on the film.

The **aperture** is an opening that lets light into the camera. The larger the aperture is, the more light enters the camera.

Figure 2 How Refracting and Reflecting Telescopes Work

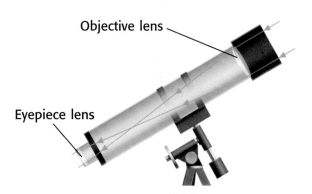

Objective lens

Eyepiece lens

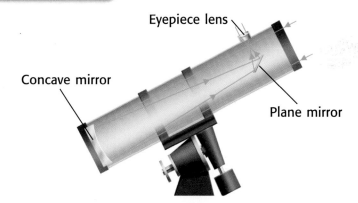

Eyepiece lens

Concave mirror

Plane mirror

A **refracting telescope** has two convex lenses. Light enters through the objective lens and forms a real image. This real image is then magnified by the eyepiece lens. You see this magnified image when you look through the eyepiece lens.

A **reflecting telescope** has a concave mirror that collects and focuses light to form a real image. The light strikes a plane mirror that directs the light to the convex eyepiece lens, which magnifies the real image.

Telescopes

Telescopes are used to see detailed images of large, distant objects. Astronomers use telescopes to study things in space, such as the moon, planets, and stars. Telescopes are classified as either refracting or reflecting. *Refracting telescopes* use lenses to collect light. *Reflecting telescopes* use mirrors to collect light. **Figure 2** shows how these two kinds of telescopes work.

Light Microscopes

Simple light microscopes are similar to refracting telescopes. These microscopes have two convex lenses. An objective lens is close to the object being studied. An eyepiece lens is the lens you look through. Microscopes are used to see magnified images of tiny, nearby objects.

Lasers and Laser Light

A **laser** is a device that produces intense light of only one color and wavelength. Laser light is different from nonlaser light in many ways. One important difference is that laser light is *coherent*. When light is coherent, light waves move together as they travel away from their source. The crests and troughs of coherent light waves are aligned. So, the individual waves behave as one wave.

✓ Reading Check What does it mean for light to be coherent?
(*See the Appendix for answers to Reading Checks.*)

Microscope Magnification
Some microscopes use more than one lens to magnify objects. The power of each lens indicates the amount of magnification the lens gives. For example, a 10× lens magnifies objects 10 times. To find the amount of magnification given by two or more lenses used together, multiply the powers of the lenses. What is the magnification given by a 5× lens used with a 20× lens?

laser a device that produces intense light of only one wavelength and color

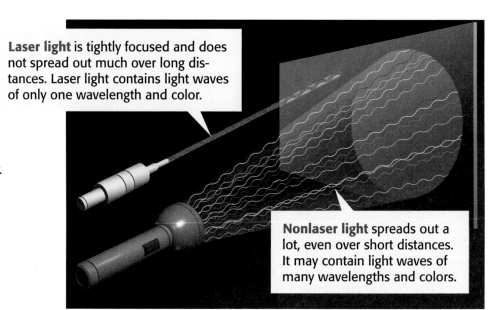

Laser light is tightly focused and does not spread out much over long distances. Laser light contains light waves of only one wavelength and color.

Nonlaser light spreads out a lot, even over short distances. It may contain light waves of many wavelengths and colors.

Figure 3 *Laser light is very different from nonlaser light.*

How Lasers Produce Light

Figure 3 compares laser and nonlaser light. The word *laser* stands for **l**ight **a**mplification by **s**timulated **e**mission of **r**adiation. *Amplification* is the increase in the brightness of the light. *Radiation* is energy transferred as electromagnetic waves.

What is stimulated emission? In an atom, an electron can move from one energy level to another. A photon (a particle of light) is released when an electron moves from a higher energy level to a lower energy level. The release of photons is called *emission*. *Stimulated emission* occurs when a photon strikes an atom that is in an excited state and makes the atom emit another photon. The newly emitted photon is identical to the first photon. The two photons travel away from the atom together. **Figure 4** shows how laser light is produced.

Figure 4 How a Helium-Neon Laser Works

a The inside of the laser is filled with helium and neon gases. An electric current in the laser excites the atoms of the gases.

b Excited neon atoms release photons of red light. When these photons strike other excited neon atoms, stimulated emission occurs.

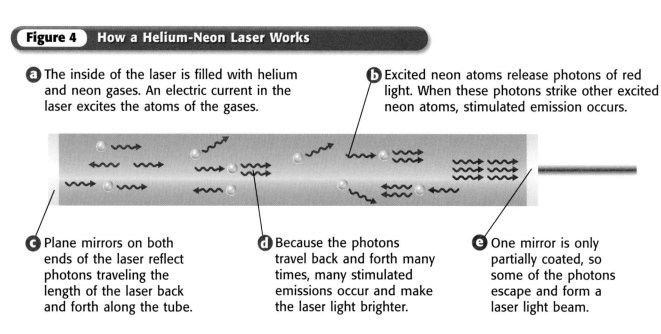

c Plane mirrors on both ends of the laser reflect photons traveling the length of the laser back and forth along the tube.

d Because the photons travel back and forth many times, many stimulated emissions occur and make the laser light brighter.

e One mirror is only partially coated, so some of the photons escape and form a laser light beam.

Uses for Lasers

Lasers are used to make holograms, such as the one shown in **Figure 5.** A **hologram** is a piece of film that produces a three-dimensional image of an object. Holograms are similar to photographs because both are images recorded on film. However, unlike photographs, the images you see in holograms are not on the surface of the film. The images appear in front of or behind the film. If you move the hologram, you will see the image from different angles.

Lasers are also used for other tasks. For example, lasers are used to cut materials such as metal and cloth. Doctors sometimes use lasers for surgery. And CD players have lasers. Light from the laser in a CD player reflects off patterns on a CD's surface. The reflected light is converted to a sound wave.

✓ Reading Check How are holograms like photographs?

Optical Fibers

Imagine a glass thread that transmits more than 1,000 telephone conversations at the same time with flashes of light. This thread, called an *optical fiber,* is a thin, glass wire that transmits light over long distances. Some optical fibers are shown in **Figure 6.** Transmitting information through telephone cables is the most common use of optical fibers. Optical fibers are also used to network computers. And they allow doctors to see inside patients' bodies without performing major surgery.

Light in a Pipe

Optical fibers are like pipes that carry light. Light stays inside an optical fiber because of total internal reflection. *Total internal reflection* is the complete reflection of light along the inside surface of the material through which it travels. **Figure 6** shows total internal reflection in an optical fiber.

hologram a piece of film that produces a three-dimensional image of an object; made by using laser light

Figure 5 *Some holograms make three-dimensional images that look so real that you might want to reach out and touch them!*

| **Figure 6** | **How Optical Fibers Work** |

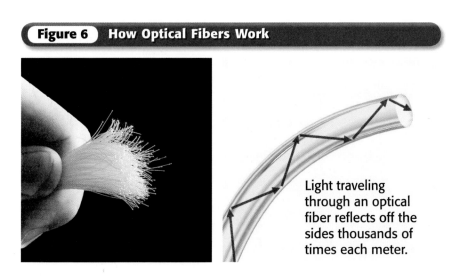

Light traveling through an optical fiber reflects off the sides thousands of times each meter.

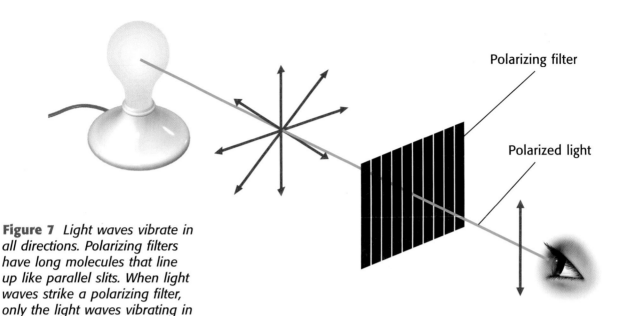

Polarizing filter

Polarized light

Figure 7 *Light waves vibrate in all directions. Polarizing filters have long molecules that line up like parallel slits. When light waves strike a polarizing filter, only the light waves vibrating in the same direction as the slits pass through.*

Polarized Light

The next time you shop for sunglasses, look for some that have lenses that polarize light. Such sunglasses are good for reducing glare. *Polarized light* consists of light waves that vibrate in only one plane. **Figure 7** illustrates how light is polarized.

When light reflects off a horizontal surface, such as a car hood or a body of water, the light is polarized horizontally. You see this polarized light as glare. Polarizing sunglasses reduce glare from horizontal surfaces because the lenses have vertically polarized filters. These filters allow only vertically vibrating light waves to pass through them. Polarizing filters are also used by photographers to reduce glare in their photographs, as shown in **Figure 8.**

Figure 8 *These two photos were taken by the same camera and from the same angle. There is less reflected light in the photo at right because a polarizing filter was placed over the lens of the camera.*

Quick Lab

Blackout!

1. Hold a **lens from a pair of polarizing sunglasses** up to your eye, and look through the lens. Record your observations.

2. Put a **second polarizing lens** over the first lens. Make sure both lenses are right side up. Look through both lenses, and describe your observations.

3. Rotate one lens slowly as you look through both lenses, and describe what happens.

4. Why can't you see through the lenses when they are aligned a certain way?

Communication Technology

You may think that talking on the telephone has nothing to do with light. But if you are talking on a cordless telephone or a cellular telephone, you are using a form of light technology! Light is an electromagnetic wave. There are many different kinds of electromagnetic waves. Radio waves and microwaves are kinds of electromagnetic waves. And cordless telephones and cellular telephones use radio waves and microwaves to send signals.

Cordless Telephones

Cordless telephones are a combination of a regular telephone and a radio. There are two parts to a cordless telephone—the base and the handset. The base is connected to a telephone jack in the wall of a building. The base receives calls through the phone line. The base then changes the signal to a radio wave and sends the signal to the handset. The handset changes the radio signal to sound for you to hear. The handset also changes your voice to a radio wave that is sent back to the base.

Reading Check What kind of electromagnetic wave does a cordless telephone use?

Figure 9 *You can make and receive calls with a cellular telephone almost everywhere you go.*

Cellular Telephones

The telephone in **Figure 9** is a cellular telephone. Cellular telephones are similar to the handset part of a cordless telephone because they send and receive signals. But a cellular telephone receives signals from tower antennas located across the country instead of from a base. And instead of using radio waves, cellular telephones use microwaves to send information.

Satellite Television

Another technology that uses electromagnetic waves to transmit data is satellite television. Satellite television companies broadcast microwave signals from human-made satellites in space. Broadcasting from space allows more people to receive the signals than broadcasting from an antenna on Earth. Small satellite dishes on the roofs of houses or outside apartments collect the signals. The signals are then sent to the customer's television set. People who have satellite television usually have better TV reception than people who receive broadcasts from antennas on Earth.

The Global Positioning System

The Global Positioning System (GPS) is a network of 27 satellites that orbit Earth. These satellites continuously send microwave signals. The signals can be picked up by a GPS receiver on Earth and used to measure positions on the Earth's surface. **Figure 10** explains how GPS works. GPS was originally used by the United States military. But now, anyone in the world who has a GPS receiver can use the system. People use GPS to avoid getting lost and to have fun. Some cars have GPS road maps that can tell the car's driver how to get to a certain place. Hikers and campers use GPS receivers to find their way in the wilderness. And some people use GPS receivers for treasure-hunt games.

Reading Check What are two uses for GPS?

CONNECTION TO
Social Studies

Navigation GPS is a complex navigation system. Before GPS was developed, travelers and explorers used other techniques, such as compasses and stars, to find their way. Research an older form of navigation, and make a poster that summarizes what you learn.

ACTIVITY

Figure 10 The Global Positioning System

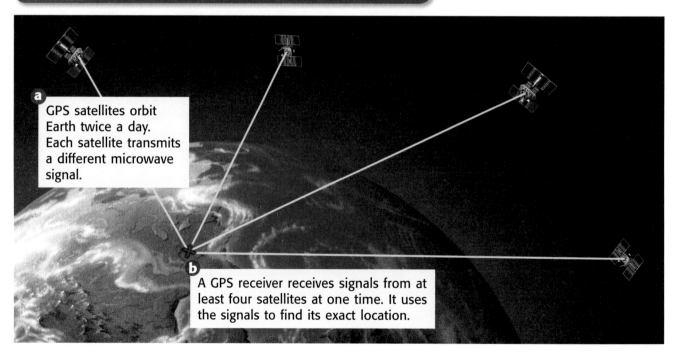

a GPS satellites orbit Earth twice a day. Each satellite transmits a different microwave signal.

b A GPS receiver receives signals from at least four satellites at one time. It uses the signals to find its exact location.

Summary

- Optical instruments, such as cameras, telescopes, and microscopes, are devices that help people make observations.
- Lasers are devices that produce intense, coherent light of only one wavelength and color. Lasers produce light by a process called *stimulated emission*.
- Optical fibers transmit light over long distances.
- Polarized light contains light waves that vibrate in only one direction.

- Cordless telephones are a combination of a telephone and a radio. Information is transmitted in the form of radio waves between the handset and the base.
- Cellular phones transmit information in the form of microwaves to and from antennas.
- Satellite television is broadcast by microwaves from satellites in space.
- GPS is a navigation system that uses microwave signals sent by a network of satellites in space.

Using Key Terms

1. Use each of the following terms in a separate sentence: *laser* and *hologram*.

Understanding Key Ideas

2. Which of the following statements about laser light is NOT true?
   a. Laser light is coherent.
   b. Laser light contains light of only one wavelength.
   c. Laser light is produced by stimulated emission.
   d. Laser light spreads out over short distances.

3. List three optical instruments, and describe what they do.

4. What are four uses for lasers?

5. Describe how optical fibers work.

6. What is polarized light?

7. Describe two ways that satellites in space are useful in everyday life.

Critical Thinking

8. **Making Comparisons** Compare how a cordless telephone works with how a cellular telephone works.

9. **Making Inferences** Why do you think optical fibers can transmit information over long distances without losing much of the signal?

Interpreting Graphics

Use the graph below to answer the questions that follow.

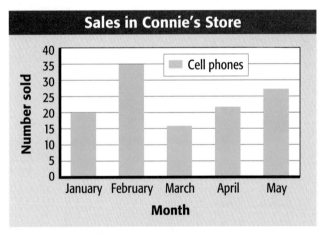

Sales in Connie's Store

10. In which two months did Connie's store sell the most cellular telephones?

11. How many cellular telephones were sold in January?

Skills Practice Lab

Images from Convex Lenses

A convex lens is thicker in the center than at the edges. Light rays passing through a convex lens come together at a focal point. Under certain conditions, a convex lens will create a real image of an object. This image will have certain characteristics, depending on the distance between the object and the lens. In this experiment, you will determine the characteristics of real images created by a convex lens—the kind of lens used as a magnifying lens.

OBJECTIVES

Use a convex lens to form images.

Determine the characteristics of real images formed by convex lenses.

MATERIALS

- candle
- card, index, 4 × 6 in. or larger
- clay, modeling
- convex lens
- jar lid
- matches
- meterstick

SAFETY

Ask a Question

1 What are the characteristics of real images created by a convex lens? For example, are the images upright or inverted (upside down)? Are the images larger or smaller than the object?

Form a Hypothesis

2 Write a hypothesis that is a possible answer to the questions above. Explain your reasoning.

Test the Hypothesis

3 Copy the table below.

	Data Collection			
Image	Orientation (upright/ inverted)	Size (larger/ smaller)	Image distance (cm)	Object distance (cm)
1				
2	DO NOT WRITE IN BOOK			
3				

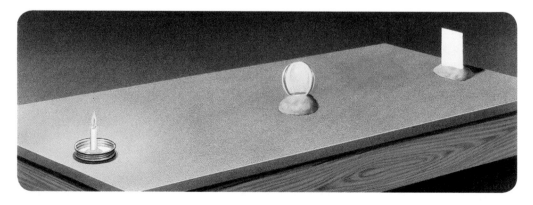

4 Use modeling clay to make a base for the lens. Place the lens and base in the middle of the table.

5 Stand the index card upright in some modeling clay on one side of the lens.

6 Place the candle in the jar lid, and anchor it with some modeling clay. Place the candle on the table so that the lens is halfway between the candle and the card. Light the candle.
Caution: Use extreme care around an open flame.

7 In a darkened room, slowly move the card and the candle away from the lens while keeping the lens exactly halfway between the card and the candle. Continue until you see a clear image of the candle flame on the card. This is image 1.

8 Measure and record the distance between the lens and the card (image distance) and between the lens and the candle (object distance).

9 Is the image upright or inverted? Is it larger or smaller than the candle? Record this information in the table.

10 Move the lens toward the candle. The new object distance should be less than half the object distance measured in step 8. Move the card back and forth until you find a sharp image (image 2) of the candle on the card.

11 Repeat steps 8 and 9 for image 2.

12 Leave the card and candle in place and move the lens toward the card to get the third image (image 3).

13 Repeat steps 8 and 9 for image 3.

Analyze the Results

1 **Recognizing Patterns** Describe the trend between image distance and image size.

2 **Examining Data** What are the similarities between the real images that are formed by a convex lens?

Draw Conclusions

3 **Making Predictions** The lens of your eye is a convex lens. Use the information you collected to describe the image projected on the back of your eye when you look at an object.

Applying Your Data

Convex lenses are used in film projectors. Explain why your favorite movie stars are truly "larger than life" on the screen in terms of image distance and object distance.

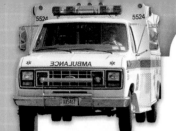

Chapter Review

USING KEY TERMS

In each of the following sentences, replace the incorrect term with the correct term from the word bank.

nearsightedness   hologram
concave mirror   laser
plane mirror   convex lens
convex mirror   farsightedness

1 A convex mirror is a mirror shaped like the inside of a spoon.

2 Eye surgeons use a hologram to reshape the cornea of an eye.

3 A person who has nearsightedness has trouble reading a book.

4 A concave lens refracts light and focuses it inward to a focal point.

5 If you move a lens around, you can see its three-dimensional image from different angles.

UNDERSTANDING KEY IDEAS

Multiple Choice

6 Which of the following parts of the eye refracts light?

a. pupil   **c.** lens
b. iris   **d.** retina

7 A vision problem that happens when light is focused in front of the retina is

a. farsightedness.
b. nearsightedness.
c. color deficiency.
d. None of the above

8 What kind of mirror provides images of large areas and is used for security?

a. a plane mirror
b. a concave mirror
c. a convex mirror
d. All of the above

9 A simple refracting telescope has

a. a convex lens and a concave lens.
b. a concave mirror and a convex lens.
c. two convex lenses.
d. two concave lenses.

10 Light waves in a laser beam interact and act as one wave. This light is called

a. coherent light.   **c.** polarized light.
b. emitted light.   **d.** reflected light.

11 When you look at yourself in a plane mirror, you see a

a. real image behind the mirror.
b. real image on the surface of the mirror.
c. virtual image that appears to be behind the mirror.
d. virtual image that appears to be in front of the mirror.

Short Answer

12 What kind of eyeglass lens should be prescribed for a person who cannot focus on nearby objects? Explain.

13 How is a hologram different from a photograph?

14 Why might a scientist who is working at the North Pole need polarizing sunglasses?

Math Skills

15 Ms. Welch's class conducted a poll about vision problems. Of the 150 students asked, 21 reported that they are nearsighted. Six of the nearsighted students wear contact lenses to correct their vision, and the rest wear glasses.

a. What percentage of the students asked is nearsighted?

b. What percentage of the students asked wears glasses?

CRITICAL THINKING

16 Concept Mapping Use the following terms to create a concept map: *lens, telescope, camera, real image, virtual image,* and *optical instrument.*

17 Analyzing Ideas Stoplights are usually mounted so that the red light is on the top and the green light is on the bottom. Why is it important for a person who has red-green color deficiency to know this arrangement?

18 Applying Concepts How could you find out if a device that produces red light is a laser or if it is just a red flashlight?

19 Making Inferences Imagine that you have a GPS receiver. When you use your receiver in the park and are surrounded by tall trees, the receiver easily finds your location. But when you use your receiver downtown and are surrounded by tall buildings, the receiver cannot determine your location. Why do you think there is a difference in reception? Describe a situation in which poor GPS reception around tall buildings could cause problems.

INTERPRETING GRAPHICS

20 Look at the ray diagrams below. For each diagram, identify the type of mirror that is being used and the kind of image that is being formed.

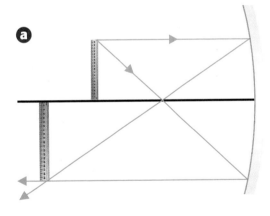

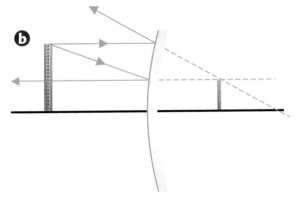

Standardized Test Preparation

Read each of the passages below. Then, answer the questions that follow each passage.

Passage 1 One day in the 1920s, an automobile collided with a horse and carriage. Garrett Morgan witnessed this, and the accident gave him an idea. Morgan designed a signal that included signs to direct traffic at busy intersections. The signal could be seen from a distance and could be clearly understood. Morgan patented the first traffic signal in 1923. Unlike the small, three-bulb signal boxes used today, the early <u>versions</u> were T shaped and had the words *stop* and *go* printed on them.

Morgan's invention was an immediate success. Morgan sold the patent to General Electric Corporation for $40,000—a large sum in those days. Since then, later versions of Morgan's traffic signal have been a mainstay of traffic control.

1. In the passage, what does the word *versions* refer to?

A automobiles

B accidents

C light bulbs

D traffic signals

2. Which of the following statements is a fact?

F Morgan still makes money selling traffic signals today.

G Traffic signals were confusing and caused a lot of accidents.

H Morgan came up with the idea of a traffic signal after seeing a traffic accident.

I Morgan patented the traffic signal in 1920.

3. How were the first traffic signals similar to the signals used today?

A They were T shaped.

B They contained three light bulbs.

C The words *stop* and *go* were printed on them.

D They directed traffic at busy intersections.

Passage 2 Twenty years ago, stars were very visible, even above large cities. Now, the stars above large cities are <u>obscured</u> by the glow from city lights. This glow, called *sky glow,* is created when light reflects off dust and particles in the atmosphere. Sky glow is also called *light pollution.*

The majority of light pollution comes from outdoor lights, such as headlights, street lights, porch lights, and parking-lot lights. Unlike other kinds of pollution, light pollution can easily be reduced. For example, using covered outdoor lights keeps the light angled downward, which prevents most of the light from reaching particles in the sky. Also, using motion-sensitive lights and timed lights helps eliminate unnecessary light.

1. Which of the following **best** describes the reason the author wrote the passage?

A to explain light pollution and to explain how to reduce it

B to convince people to look at stars

C to explain why people should not live in cities

D to describe the beauty of sky glow

2. Which of the following contributes the least amount to light pollution?

F headlights on cars

G lights inside homes

H lights used in outdoor stadiums

I lights in large parking lots

3. In the passage, what does the word *obscured* mean?

A made brighter

B reflected

C polluted

D made difficult to see

The table below shows details about four lasers sold by a laser company. Use the table below to answer the questions that follow.

Laser Specifications			
Color	Power (mW)	Wavelength (nm)	Mass (kg)
Blue	15	488	2.8
Yellow	5	568	5.8
Red	18	633	0.9
Red	10	633	0.6

1. What is the mass of the laser that has the most power?
   A 0.6 kg
   B 0.9 kg
   C 2.8 kg
   D 5.8 kg

2. The company also sells a laser that has a wavelength of 633 nm and a power of 5 mW. Which of the following statements **best** predicts the mass of this laser?
   F The laser has a mass of less than 0.6 kg.
   G The laser has a mass between 0.6 kg and 0.9 kg.
   H The laser has a mass greater than 0.9 kg.
   I The laser has a mass of 5.8 kg.

3. Based on the information in the table, which statement is most likely true?
   A The power of the laser determines the color of light.
   B The wavelength of the laser determines the color of light.
   C The mass of the laser determines the color of light.
   D There is not enough information to determine the answer.

Read each question below, and choose the best answer.

1. Micah has a box that has a length of 16 cm, a width of 10 cm, and a height of 5 cm. What is the volume of the box?
   A 1,600 cm^3
   B 800 cm^3
   C 700 cm^3
   D 500 cm^3

2. The table below shows the low temperature in Minneapolis, Minnesota, for five days in December.

Day	Temperature (°C)
Monday	−12
Tuesday	−8
Wednesday	7
Thursday	−3
Friday	11

Which list shows the temperatures from lowest to highest?
   F −3°C, −8°C, −12°C, 7°C, 11°C
   G −3°C, 7°C, −8°C, 11°C, −12°C
   H −12°C, 11°C, 7°C, −8°C, −3°C
   I −12°C, −8°C, −3°C, 7°C, 11°C

3. The power of a microscope lens is the amount of magnification the lens gives. For example, a 10× lens magnifies objects 10 times. How many times is an object magnified if it is viewed with both a 5× lens and a 30× lens?
   A 35 times
   B 60 times
   C 150 times
   D 350 times

Science in Action

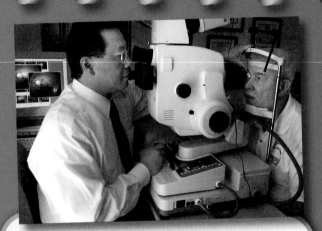

Science, Technology, and Society

Bionic Eyes

Imagine bionic eyes that allow a person who is blind to see. Researchers working on artificial vision think that the technology will be available soon. Many companies are working on different ways to restore sight to people who are blind. Some companies are developing artificial corneas, while other companies are building artificial retinas. One item that has already been tested on people is a pair of glasses that provides limited vision. The glasses have a camera that sends a signal to an electrode implanted in the person's brain. The images are black and white and are not detailed, but the person who is wearing the glasses can see obstacles in his or her path.

Language Arts ACTiViTY

WRITING SKILL Write a one-page story about a teen who has his or her eyesight restored by a bionic eye. What would the teen want to see first? What would the teen do that he or she couldn't do before?

Scientific Debate

Do Cellular Telephones Cause Cancer?

As cellular telephones became popular, people began to wonder if the phones were dangerous. Some cell-phone users claimed that the microwave energy from their cell phones caused them to develop brain cancer. So far, most research shows that the microwave energy emitted by cell phones is too low and too weak to damage human tissue. However, some studies have shown negative effects. There is some evidence that the low-power microwave energy used by cell phones may damage DNA and may cause cells to shrink. Because so many people use cellular phones, research continues around the world.

Math ACTiViTY

The American Cancer Society estimates that 0.006% of people in the United States will be diagnosed with brain cancer each year. If a city has a population of 50,000 people, how many people in that city will be diagnosed with brain cancer in one year?

Sandra Faber

Astronomer What do you do when you send a telescope into space and then find out that it is broken? You call Dr. Sandra Faber, a professor of astronomy at the University of California, Santa Cruz (UCSC). In April 1990, after the *Hubble Space Telescope* went into orbit, scientists found that the images the telescope collected were not turning out as expected. Dr. Faber's team at UCSC was in charge of a device on *Hubble* called the *Wide Field Planetary Camera*. Dr. Faber and her team decided to test the telescope to determine what was wrong.

To perform the test, they centered *Hubble* onto a bright star and took several photos. From those photos, Dr. Faber's team created a model of what was wrong. After reporting the error to NASA and presenting the model they had developed, Dr. Faber and a group of experts began to correct the problem. The group's efforts were a success and put *Hubble* back into operation so that astronomers could continue researching stars and other objects in space.

Social Studies ACTiViTY

Research the history of the telescope. Make a timeline with the dates of major events in telescope history. For example, you could include the first use of a telescope to see the rings of Saturn in your timeline.

To learn more about these Science in Action topics, visit **go.hrw.com** and type in the keyword **HP5LOWF.**

Current Science

Check out Current Science® articles related to this chapter by visiting go.hrw.com. Just type in the keyword **HP5CS23.**

Skills Practice Lab

Wave Speed, Frequency, and Wavelength

Wave speed, frequency, and wavelength are three related properties of waves. In this lab, you will make observations and collect data to determine the relationship among these properties.

MATERIALS

- meterstick
- stopwatch
- toy, coiled spring

SAFETY

Part A: Wave Speed

Procedure

1. Copy Table 1.

Table 1 Wave Speed Data			
Trial	Length of spring (m)	Time for wave (s)	Speed of wave (m/s)
1			
2		DO NOT WRITE IN BOOK	
3			
Average			

2. Two students should stretch the spring to a length of 2 m to 4 m on the floor or on a table. A third student should measure the length of the spring. Record the length in Table 1.

3. One student should pull part of the spring sideways with one hand, as shown at right, and release the pulled-back portion. This action will cause a wave to travel down the spring.

4. Using a stopwatch, the third student should measure how long it takes for the wave to travel down the length of the spring and back. Record this time in Table 1.

5. Repeat steps 3 and 4 two more times.

Part B: Wavelength and Frequency

Procedure

1️⃣ Keep the spring the same length that you used in Part A.

2️⃣ Copy Table 2.

Table 2 Wavelength and Frequency Data				
Trial	Length of spring (m)	Time for 10 cycles (s)	Wave frequency (Hz)	Wavelength (m)
1				
2		*DO NOT WRITE IN BOOK*		
3				
Average				

3️⃣ One of the two students holding the spring should start shaking the spring from side to side until a wave pattern appears that resembles one of those shown.

4️⃣ Using the stopwatch, the third student should measure and record how long it takes for 10 cycles of the wave pattern to occur. (One back-and-forth shake is 1 cycle.) Keep the pattern going so that measurements for three trials can be made.

Analyze the Results

Part A

1️⃣ Calculate and record the wave speed for each trial. (Speed equals distance divided by time; distance is twice the spring length.)

2️⃣ Calculate and record the average time and the average wave speed.

Part B

3️⃣ Calculate the frequency for each trial by dividing the number of cycles (10) by the time. Record the answers in Table 2.

4️⃣ Determine the wavelength using the equation at right that matches your wave pattern. Record your answer in Table 2.

5️⃣ Calculate and record the average time and frequency.

Draw Conclusions: Parts A and B

6️⃣ Analyze the relationship among speed, wavelength, and frequency. Multiply or divide any two of them to see if the result equals the third. (Use the averages from your data tables.) Write the equation that shows the relationship.

Wave Patterns

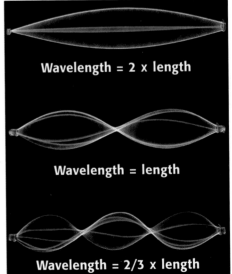

Wavelength = 2 x length

Wavelength = length

Wavelength = 2/3 x length

Inquiry Lab

The Speed of Sound

In the chapter entitled "The Nature of Sound," you learned that the speed of sound in air is 343 m/s at 20°C (approximately room temperature). In this lab, you'll design an experiment to measure the speed of sound yourself—and you'll determine if you're "up to speed"!

MATERIALS

- items to be determined by the students and approved by the teacher

Procedure

1 Brainstorm with your teammates to come up with a way to measure the speed of sound. Consider the following as you design your experiment:

a. You must have a method of making a sound. Some simple examples include speaking, clapping your hands, and hitting two boards together.

b. Remember that speed is equal to distance divided by time. You must devise methods to measure the distance that a sound travels and to measure the amount of time it takes for that sound to travel that distance.

c. Sound travels very rapidly. A sound from across the room will reach your ears almost before you can start recording the time! You may wish to have the sound travel a long distance.

d. Remember that sound travels in waves. Think about the interactions of sound waves. You might be able to include these interactions in your design.

2 Discuss your experimental design with your teacher, including any equipment you need. Your teacher may have questions that will help you improve your design.

3 Once your design is approved, carry out your experiment. Be sure to perform several trials. Record your results.

Analyze the Results

1 Was your result close to the value given in the introduction to this lab? If not, what factors may have caused you to get such a different value?

2 Why was it important for you to perform several trials in your experiment?

Draw Conclusions

3 Compare your results with those of your classmates. Determine which experimental design provided the best results. Explain why you think this design was so successful.

Skills Practice Lab

Tuneful Tube

If you have seen a singer shatter a crystal glass simply by singing a note, you have seen an example of resonance. For the glass to shatter, the note has to match the resonant frequency of the glass. A column of air within a cylinder can also resonate if the air column is the proper length for the frequency of the note. In this lab, you will investigate the relationship between the length of an air column, the frequency, and the wavelength during resonance.

Procedure

1 Copy the data table below.

Data Collection Table			
Frequency (Hz)			
Length (cm)	DO NOT WRITE IN BOOK		

2 Fill the graduated cylinder with water.

3 Hold a plastic tube in the water so that about 3 cm is above the water.

4 Record the frequency of the first tuning fork. Gently strike the tuning fork with the eraser, and hold the tuning fork so that the prongs are just above the tube, as shown at right. Slowly move the tube and fork up and down until you hear the loudest sound.

5 Measure the distance from the top of the tube to the water. Record this length in your data table.

6 Repeat steps 3–5 using the other three tuning forks.

Analyze the Results

1 Calculate the wavelength (in centimeters) of each sound wave by dividing the speed of sound in air (343 m/s at 20°C) by the frequency and multiplying by 100.

2 Make the following graphs: air column length versus frequency and wavelength versus frequency. On both graphs, plot the frequency on the x-axis.

3 Describe the trend between the length of the air column and the frequency of the tuning fork.

4 How are the pitches you heard related to the wavelengths of the sounds?

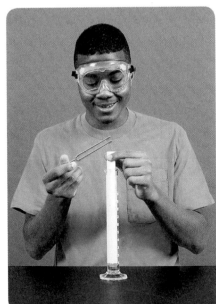

Skills Practice Lab

The Energy of Sound

In the chapter entitled "The Nature of Sound," you learned about various properties and interactions of sound. In this lab, you will perform several activities that will demonstrate that the properties and interactions of sound all depend on one thing—the energy carried by sound waves.

MATERIALS

- cup, plastic, small, filled with water
- eraser, pink, rubber
- rubber band
- string, 50 cm
- tuning forks, same frequency (2), different frequency (1)

SAFETY

Part A: Sound Vibrations

Procedure

1. Lightly strike a tuning fork with the eraser. Slowly place the prongs of the tuning fork in the plastic cup of water. Record your observations.

Part B: Resonance

Procedure

1. Strike a tuning fork with the eraser. Quickly pick up a second tuning fork in your other hand, and hold it about 30 cm from the first tuning fork.

2. Place the first tuning fork against your leg to stop the tuning fork's vibration. Listen closely to the second tuning fork. Record your observations, including the frequencies of the two tuning forks.

3. Repeat steps 1 and 2, using the remaining tuning fork as the second tuning fork.

Part C: Interference

Procedure

1. Use the two tuning forks that have the same frequency, and place a rubber band tightly over the prongs near the base of one tuning fork, as shown at right. Strike both tuning forks against the eraser. Hold the stems of the tuning forks against a table, 3 cm to 5 cm apart. If you cannot hear any differences, move the rubber band up or down the prongs. Strike again. Record your observations.

Part D: The Doppler Effect

Procedure

1 Your teacher will tie the piece of string securely to the base of one tuning fork. Your teacher will then strike the tuning fork and carefully swing the tuning fork in a circle overhead. Record your observations.

Analyze the Results

1 How do your observations demonstrate that sound waves are carried through vibrations?

2 Explain why you can hear a sound from the second tuning fork when the frequencies of the tuning forks used are the same.

3 When using tuning forks of different frequencies, would you expect to hear a sound from the second tuning fork if you strike the first tuning fork harder? Explain your reasoning.

4 Did you notice the sound changing back and forth between loud and soft? A steady pattern like this one is called a *beat frequency.* Explain this changing pattern of loudness and softness in terms of interference (both constructive and destructive).

5 Did the tuning fork make a different sound when your teacher was swinging it than when he or she was holding it? If yes, explain why.

6 Is the actual pitch of the tuning fork changing when it is swinging? Explain.

Draw Conclusions

7 Explain how your observations from each part of this lab verify that sound waves carry energy from one point to another through a vibrating medium.

8 Particularly loud thunder can cause the windows of your room to rattle. How is this evidence that sound waves carry energy?

Skills Practice Lab

What Color of Light Is Best for Green Plants?

Plants grow well outdoors under natural sunlight. However, some plants are grown indoors under artificial light. A variety of colored lights are available for helping plants grow indoors. In this experiment, you'll test several colors of light to discover which color best meets the energy needs of green plants.

Ask a Question

1 Which color of light is best for growing green plants?

Form a Hypothesis

2 Write a hypothesis that answers the question above. Explain your reasoning.

Test the Hypothesis

3 Use the masking tape and marker to label the side of each Petri dish with your name and the type of light under which you will place the dish.

4 Place a moist paper towel in each Petri dish. Place 5 seedlings on top of the paper towel. Cover each dish.

5 Record your observations of the seedlings, such as length, color, and number of leaves.

6 Place each dish under the appropriate light.

7 Observe the Petri dishes every day for at least 5 days. Record your observations.

Analyze the Results

1 Based on your results, which color of light is best for growing green plants? Which color of light is worst?

Draw Conclusions

2 Remember that the color of an opaque object (such as a plant) is determined by the colors the object reflects. Use this information to explain your answer to question 1 above.

3 Would a purple light be good for growing purple plants? Explain.

MATERIALS

- bean seedlings
- colored lights, supplied by your teacher
- marker, felt-tip
- paper towels
- Petri dishes with covers
- tape, masking
- water

SAFETY

Skills Practice Lab

Which Color Is Hottest?

Will a navy blue hat or a white hat keep your head warmer in cool weather? Colored objects absorb energy, which can make the objects warmer. How much energy is absorbed depends on the object's color. In this experiment, you will test several colors under a bright light to determine which colors absorb the most energy.

MATERIALS

- light source
- paper, colored, squares
- paper, graph
- paper towels
- pencils or pens, colored
- tape, transparent
- thermometer
- water, room-temperature

SAFETY

Procedure

1. Copy the table below. Be sure to have one column for each color of paper you use and enough rows to end at 3 min.

Data Collection Table				
Time (s)	White	Red	Blue	Black
0				
15				
30		DO NOT WRITE IN BOOK		
45				
etc.				

2. Tape a piece of colored paper around the bottom of a thermometer, and hold it under the light source. Record the temperature every 15 s for 3 min.

3. Cool the thermometer by removing the piece of paper and placing the thermometer in the cup of room-temperature water. After 1 min, remove the thermometer, and dry it with a paper towel.

4. Repeat steps 2 and 3 with each color, making sure to hold the thermometer at the same distance from the light source.

Analyze the Results

1. Prepare a graph of temperature (*y*-axis) versus time (*x*-axis). Using a different colored pencil or pen for each set of data, plot all data on one graph.

2. Rank the colors you used in order from hottest to coolest.

Draw Conclusions

3. Compare the colors, based on the amount of energy each absorbs.

4. In this experiment, a white light was used. How would your results be different if you used a red light? Explain.

5. Use the relationship between color and energy absorbed to explain why different colors of clothing are used for different seasons.

Skills Practice Lab

Mirror Images

When light actually passes through an image, the image is a real image. When light does not pass through the image, the image is a virtual image. Recall that plane mirrors produce only virtual images because the image appears to be behind the mirror where no light can pass through it.

In fact, all mirrors can form virtual images, but only some mirrors can form real images. In this experiment, you will explore the virtual images formed by concave and convex mirrors, and you will try to find a real image using both types of mirrors.

MATERIALS

- candle
- card, index
- clay, modeling
- jar lid
- matches
- mirror, concave
- mirror, convex

SAFETY

Part A: Finding Virtual Images

Procedure

1. Hold the convex mirror at arm's length away from your face. Observe the image of your face in the mirror.

2. Slowly move the mirror toward your face, and observe what happens to the image. Record your observations.

3. Move the mirror very close to your face. Record your observations.

4. Slowly move the mirror away from your face, and observe what happens to the image. Record your observations.

5. Repeat steps 1 through 4 with the concave mirror.

Analyze the Results

1. For each mirror, did you find a virtual image? How can you tell?

2. Describe the images you found. Were they smaller than, larger than, or the same size as your face? Were they right side up or upside down?

Draw Conclusions

3. Describe at least one use for each type of mirror. Be creative, and try to think of inventions that might use the properties of the two types of mirrors.

Part B: Finding a Real Image

Procedure

1 In a darkened room, place a candle in a jar lid near one end of a table. Use modeling clay to hold the candle in place. Light the candle. **Caution:** Use extreme care around an open flame.

2 Use more modeling clay to make a base to hold the convex mirror upright. Place the mirror at the other end of the table, facing the candle.

3 Hold the index card between the candle and the mirror but slightly to one side so that you do not block the candlelight, as shown below.

4 Move the card slowly from side to side and back and forth to see whether you can focus an image of the candle on it. Record your results.

5 Repeat steps 2–4 with the concave mirror.

Analyze the Results

1 For each mirror, did you find a real image? How can you tell?

2 Describe the real image you found. Was it smaller than, larger than, or the same size as the object? Was it right side up or upside down?

Draw Conclusions

3 Astronomical telescopes use large mirrors to reflect light to form a real image. Based on your results, do you think a concave or a convex mirror would be better for this instrument? Explain your answer.

Contents

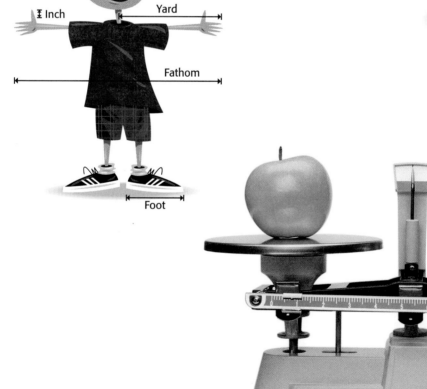

Inch

Yard

Fathom

Foot

Appendix

✓ *Reading Check* Answers

Chapter 1 The Energy of Waves
Section 1
Page 4: All waves are disturbances that transmit energy.
Page 6: Electromagnetic waves do not require a medium.
Page 8: A sound wave is a longitudinal wave.
Section 2
Page 11: Shaking the rope faster makes the wavelength shorter; shaking the rope more slowly makes the wavelength longer.
Page 12: 3 Hz
Section 3
Page 15: It refracts.
Page 17: Constructive interference occurs when the crests of one wave overlap the crests of another wave.
Page 18: A standing wave results from a wave that is reflected between two fixed points. Interference from the wave and reflected waves cause certain points to remain at rest and certain points to remain at a large amplitude.

Chapter 2 The Nature of Sound
Section 1
Page 31: Sound waves consist of longitudinal waves carried through a medium.
Page 32: Sound needs a medium in order to travel.
Page 34: Tinnitus is caused by long-term exposure to loud sounds.
Section 2
Page 37: Frequency is the number of crests or troughs made in a given time.
Page 39: The amplitude of a sound increases as the energy of the vibrations that caused the sound increases.
Page 40: An oscilloscope turns sounds into electrical signals and graphs the signals.
Section 3
Page 43: Echolocation helps some animals find food.
Page 44: Sound wave interference can be either constructive or destructive.
Page 46: A standing wave is a pattern of vibration that looks like a wave that is standing still.
Section 4
Page 49: Musical instruments differ in the part of the instrument that vibrates and in the way that the vibrations are made.
Page 51: Music consists of sound waves that have regular patterns, and noise consists of a random mix of frequencies.

Chapter 3 The Nature of Light
Section 1
Page 63: Electric fields can be found around every charged object.
Page 64: The speed of light is about 880,000 times faster than the speed of sound.
Section 2
Page 66: The speed of a wave is determined by multiplying the wavelength and frequency of the wave.
Page 68: Radio waves carry TV signals.
Page 70: White light is the combination of visible light of all wavelengths.
Page 71: Ultraviolet light waves have shorter wavelengths and higher frequencies than visible light waves do.
Page 72: Patients are protected from X rays by special lead-lined aprons.
Section 3
Page 74: The law of reflection states that the angle of incidence equals the angle of reflection.
Page 75: Sample answer: Four light sources are a television screen, a fluorescent light in the classroom, a light bulb, and the tail of a firefly.
Page 76: You can see things outside of a beam of light because light is scattered outside of the beam.
Page 79: The amount that a wave diffracts depends on the wavelength of the wave and the size of the barrier or opening.
Page 80: Constructive interference is interference in which the resulting wave has a greater amplitude than the original waves had.
Section 4
Page 83: Sample answer: Two translucent objects are a frosted window and wax paper.
Page 84: When white light shines on a colored opaque object, some of the colors of light are absorbed and some are reflected.
Page 86: A pigment is a material that gives color to a substance by absorbing some colors of light and reflecting others.

Chapter 4 Light and Our World
Section 1
Page 99: A virtual image is an image through which light does not travel.
Page 100: A concave mirror can be used to make a powerful beam of light by putting a light source at the focal point of the mirror.
Page 102: A convex lens in thicker in the middle than it is at the edges.

Section 2

Page 105: Nearsightedness happens when a person's eye is too long. Farsightedness happens when a person's eye is too short.

Page 106: The three kinds of cones are red, blue, and green.

Section 3

Page 109: When light is coherent, light waves move together as they travel away from their source. Individual waves behave as one wave.

Page 111: Holograms are like photographs because both are images recorded on film.

Page 113: A cordless telephone sends signals by using radio waves.

Page 114: Sample answer: GPS can be used by hikers and campers to find their way in the wilderness. GPS can also be used for treasure hunt games.

Appendix

Study Skills

FoldNote Instructions

Have you ever tried to study for a test or quiz but didn't know where to start? Or have you read a chapter and found that you can remember only a few ideas? Well, FoldNotes are a fun and exciting way to help you learn and remember the ideas you encounter as you learn science!

FoldNotes are tools that you can use to organize concepts. By focusing on a few main concepts, FoldNotes help you learn and remember how the concepts fit together. They can help you see the "big picture." Below you will find instructions for building 10 different FoldNotes.

Pyramid

1. Place a sheet of paper in front of you. Fold the lower left-hand corner of the paper diagonally to the opposite edge of the paper.

2. Cut off the tab of paper created by the fold (at the top).

3. Open the paper so that it is a square. Fold the lower right-hand corner of the paper diagonally to the opposite corner to form a triangle.

4. Open the paper. The creases of the two folds will have created an X.

5. Using scissors, cut along one of the creases. Start from any corner, and stop at the center point to create two flaps. Use tape or glue to attach one of the flaps on top of the other flap.

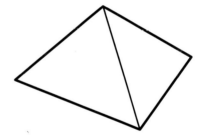

Double Door

1. Fold a sheet of paper in half from the top to the bottom. Then, unfold the paper.

2. Fold the top and bottom edges of the paper to the crease.

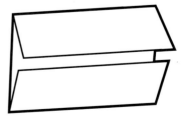

Booklet

1. Fold a sheet of paper in half from left to right. Then, unfold the paper.

2. Fold the sheet of paper in half again from the top to the bottom. Then, unfold the paper.

3. Refold the sheet of paper in half from left to right.

4. Fold the top and bottom edges to the center crease.

5. Completely unfold the paper.

6. Refold the paper from top to bottom.

7. Using scissors, cut a slit along the center crease of the sheet from the folded edge to the creases made in step 4. Do not cut the entire sheet in half.

8. Fold the sheet of paper in half from left to right. While holding the bottom and top edges of the paper, push the bottom and top edges together so that the center collapses at the center slit. Fold the four flaps to form a four-page book.

Layered Book

1. Lay one sheet of paper on top of another sheet. Slide the top sheet up so that 2 cm of the bottom sheet is showing.

2. Hold the two sheets together, fold down the top of the two sheets so that you see four 2 cm tabs along the bottom.

3. Using a stapler, staple the top of the FoldNote.

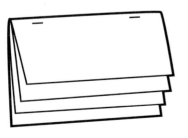

Key-Term Fold

1. Fold a sheet of lined notebook paper in half from left to right.

2. Using scissors, cut along every third line from the right edge of the paper to the center fold to make tabs.

Four-Corner Fold

1. Fold a sheet of paper in half from left to right. Then, unfold the paper.

2. Fold each side of the paper to the crease in the center of the paper.

3. Fold the paper in half from the top to the bottom. Then, unfold the paper.

4. Using scissors, cut the top flap creases made in step 3 to form four flaps.

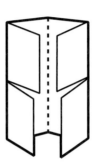

Three-Panel Flip Chart

1. Fold a piece of paper in half from the top to the bottom.

2. Fold the paper in thirds from side to side. Then, unfold the paper so that you can see the three sections.

3. From the top of the paper, cut along each of the vertical fold lines to the fold in the middle of the paper. You will now have three flaps.

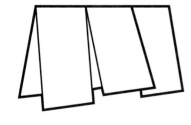

Table Fold

1. Fold a piece of paper in half from the top to the bottom. Then, fold the paper in half again.

2. Fold the paper in thirds from side to side.

3. Unfold the paper completely. Carefully trace the fold lines by using a pen or pencil.

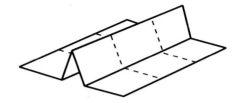

Two-Panel Flip Chart

1. Fold a piece of paper in half from the top to the bottom.

2. Fold the paper in half from side to side. Then, unfold the paper so that you can see the two sections.

3. From the top of the paper, cut along the vertical fold line to the fold in the middle of the paper. You will now have two flaps.

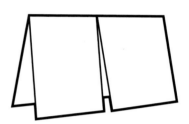

Tri-Fold

1. Fold a piece a paper in thirds from the top to the bottom.

2. Unfold the paper so that you can see the three sections. Then, turn the paper sideways so that the three sections form vertical columns.

3. Trace the fold lines by using a pen or pencil. Label the columns "Know," "Want," and "Learn."

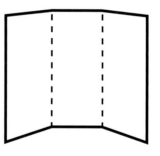

Graphic Organizer Instructions

 Have you ever wished that you could "draw out" the many concepts you learn in your science class? Sometimes, being able to *see* how concepts are related really helps you remember what you've learned. Graphic Organizers do just that! They give you a way to draw or map out concepts.

All you need to make a Graphic Organizer is a piece of paper and a pencil. Below you will find instructions for four different Graphic Organizers designed to help you organize the concepts you'll learn in this book.

Spider Map

1. Draw a diagram like the one shown. In the circle, write the main topic.

2. From the circle, draw legs to represent different categories of the main topic. You can have as many categories as you want.

3. From the category legs, draw horizontal lines. As you read the chapter, write details about each category on the horizontal lines.

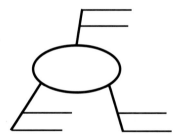

Comparison Table

1. Draw a chart like the one shown. Your chart can have as many columns and rows as you want.

2. In the top row, write the topics that you want to compare.

3. In the left column, write characteristics of the topics that you want to compare. As you read the chapter, fill in the characteristics for each topic in the appropriate boxes.

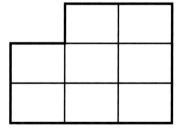

Chain-of-Events-Chart

1. Draw a box. In the box, write the first step of a process or the first event of a timeline.

2. Under the box, draw another box, and use an arrow to connect the two boxes. In the second box, write the next step of the process or the next event in the timeline.

3. Continue adding boxes until the process or timeline is finished.

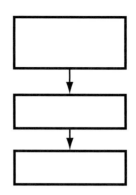

Concept Map

1. Draw a circle in the center of a piece of paper. Write the main idea of the chapter in the center of the circle.

2. From the circle, draw other circles. In those circles, write characteristics of the main idea. Draw arrows from the center circle to the circles that contain the characteristics.

3. From each circle that contains a characteristic, draw other circles. In those circles, write specific details about the characteristic. Draw arrows from each circle that contains a characteristic to the circles that contain specific details. You may draw as many circles as you want.

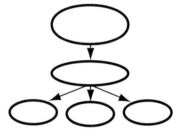

SI Measurement

The International System of Units, or SI, is the standard system of measurement used by many scientists. Using the same standards of measurement makes it easier for scientists to communicate with one another.

SI works by combining prefixes and base units. Each base unit can be used with different prefixes to define smaller and larger quantities. The table below lists common SI prefixes.

SI Prefixes

Prefix	Symbol	Factor	Example
kilo-	k	1,000	kilogram, 1 kg = 1,000 g
hecto-	h	100	hectoliter, 1 hL = 100 L
deka-	da	10	dekameter, 1 dam = 10 m
		1	meter, liter, gram
deci-	d	0.1	decigram, 1 dg = 0.1 g
centi-	c	0.01	centimeter, 1 cm = 0.01 m
milli-	m	0.001	milliliter, 1 mL = 0.001 L
micro-	μ	0.000 001	micrometer, 1 μm = 0.000 001 m

SI Conversion Table

SI units	From SI to English	From English to SI
Length		
kilometer (km) = 1,000 m	1 km = 0.621 mi	1 mi = 1.609 km
meter (m) = 100 cm	1 m = 3.281 ft	1 ft = 0.305 m
centimeter (cm) = 0.01 m	1 cm = 0.394 in.	1 in. = 2.540 cm
millimeter (mm) = 0.001 m	1 mm = 0.039 in.	
micrometer (μm) = 0.000 001 m		
nanometer (nm) = 0.000 000 001 m		
Area		
square kilometer (km^2) = 100 hectares	1 km^2 = 0.386 mi^2	1 mi^2 = 2.590 km^2
hectare (ha) = 10,000 m^2	1 ha = 2.471 acres	1 acre = 0.405 ha
square meter (m^2) = 10,000 cm^2	1 m^2 = 10.764 ft^2	1 ft^2 = 0.093 m^2
square centimeter (cm^2) = 100 mm^2	1 cm^2 = 0.155 in.^{2}	1 in.^{2} = 6.452 cm^2
Volume		
liter (L) = 1,000 mL = 1 dm^3	1 L = 1.057 fl qt	1 fl qt = 0.946 L
milliliter (mL) = 0.001 L = 1 cm^3	1 mL = 0.034 fl oz	1 fl oz = 29.574 mL
microliter (μL) = 0.000 001 L		
Mass		*Equivalent weight at Earth's surface
kilogram (kg) = 1,000 g	1 kg = 2.205 lb*	1 lb* = 0.454 kg
gram (g) = 1,000 mg	1 g = 0.035 oz*	1 oz* = 28.350 g
milligram (mg) = 0.001 g		
microgram (μg) = 0.000 001 g		

Appendix

Temperature Scales

Temperature can be expressed by using three different scales: Fahrenheit, Celsius, and Kelvin. The SI unit for temperature is the kelvin (K).

Although 0 K is much colder than 0°C, a change of 1 K is equal to a change of 1°C.

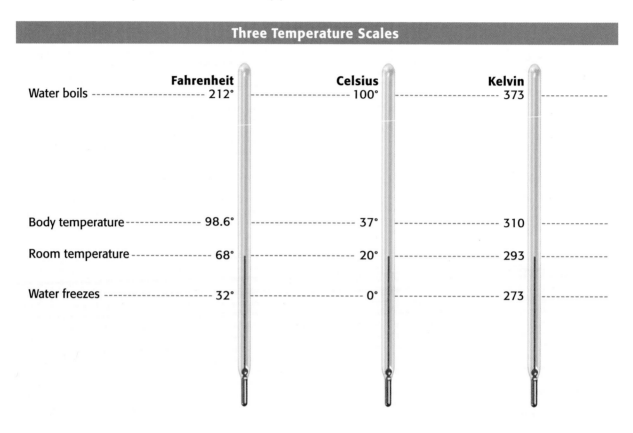

Three Temperature Scales

	Fahrenheit	Celsius	Kelvin
Water boils	212°	100°	373
Body temperature	98.6°	37°	310
Room temperature	68°	20°	293
Water freezes	32°	0°	273

Temperature Conversions Table

To convert	Use this equation:	Example
Celsius to Fahrenheit °C → °F	$°F = \left(\dfrac{9}{5} \times °C\right) + 32$	Convert 45°C to °F. $°F = \left(\dfrac{9}{5} \times 45°C\right) + 32 = 113°F$
Fahrenheit to Celsius °F → °C	$°C = \dfrac{5}{9} \times (°F - 32)$	Convert 68°F to °C. $°C = \dfrac{5}{9} \times (68°F - 32) = 20°C$
Celsius to Kelvin °C → K	$K = °C + 273$	Convert 45°C to K. $K = 45°C + 273 = 318\ K$
Kelvin to Celsius K → °C	$°C = K - 273$	Convert 32 K to °C. $°C = 32K - 273 = -241°C$

Measuring Skills

Using a Graduated Cylinder

When using a graduated cylinder to measure volume, keep the following procedures in mind:

1. Place the cylinder on a flat, level surface before measuring liquid.

2. Move your head so that your eye is level with the surface of the liquid.

3. Read the mark closest to the liquid level. On glass graduated cylinders, read the mark closest to the center of the curve in the liquid's surface.

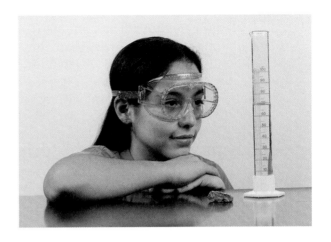

Using a Meterstick or Metric Ruler

When using a meterstick or metric ruler to measure length, keep the following procedures in mind:

1. Place the ruler firmly against the object that you are measuring.

2. Align one edge of the object exactly with the 0 end of the ruler.

3. Look at the other edge of the object to see which of the marks on the ruler is closest to that edge. (Note: Each small slash between the centimeters represents a millimeter, which is one-tenth of a centimeter.)

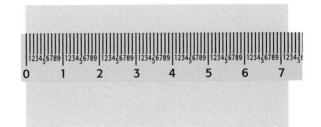

Using a Triple-Beam Balance

When using a triple-beam balance to measure mass, keep the following procedures in mind:

1. Make sure the balance is on a level surface.

2. Place all of the countermasses at 0. Adjust the balancing knob until the pointer rests at 0.

3. Place the object you wish to measure on the pan. **Caution:** Do not place hot objects or chemicals directly on the balance pan.

4. Move the largest countermass along the beam to the right until it is at the last notch that does not tip the balance. Follow the same procedure with the next-largest countermass. Then, move the smallest countermass until the pointer rests at 0.

5. Add the readings from the three beams together to determine the mass of the object.

6. When determining the mass of crystals or powders, first find the mass of a piece of filter paper. Then, add the crystals or powder to the paper, and remeasure. The actual mass of the crystals or powder is the total mass minus the mass of the paper. When finding the mass of liquids, first find the mass of the empty container. Then, find the combined mass of the liquid and container. The mass of the liquid is the total mass minus the mass of the container.

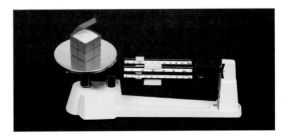

Scientific Methods

The ways in which scientists answer questions and solve problems are called **scientific methods.** The same steps are often used by scientists as they look for answers. However, there is more than one way to use these steps. Scientists may use all of the steps or just some of the steps during an investigation. They may even repeat some of the steps. The goal of using scientific methods is to come up with reliable answers and solutions.

Six Steps of Scientific Methods

1 Ask a Question

Good questions come from careful **observations.** You make observations by using your senses to gather information. Sometimes, you may use instruments, such as microscopes and telescopes, to extend the range of your senses. As you observe the natural world, you will discover that you have many more questions than answers. These questions drive investigations.

Questions beginning with *what, why, how,* and *when* are important in focusing an investigation. Here is an example of a question that could lead to an investigation.

Question: How does acid rain affect plant growth?

2 Form a Hypothesis

After you ask a question, you need to form a **hypothesis.** A hypothesis is a clear statement of what you expect the answer to your question to be. Your hypothesis will represent your best "educated guess" based on what you have observed and what you already know. A good hypothesis is testable. Otherwise, the investigation can go no further. Here is a hypothesis based on the question, "How does acid rain affect plant growth?"

Hypothesis: Acid rain slows plant growth.

The hypothesis can lead to predictions. A prediction is what you think the outcome of your experiment or data collection will be. Predictions are usually stated in an if-then format. Here is a sample prediction for the hypothesis that acid rain slows plant growth.

Prediction: If a plant is watered with only acid rain (which has a pH of 4), then the plant will grow at half its normal rate.

3 Test the Hypothesis

After you have formed a hypothesis and made a prediction, your hypothesis should be tested. One way to test a hypothesis is with a controlled experiment. A **controlled experiment** tests only one factor at a time. In an experiment to test the effect of acid rain on plant growth, the **control group** would be watered with normal rain water. The **experimental group** would be watered with acid rain. All of the plants should receive the same amount of sunlight and water each day. The air temperature should be the same for all groups. However, the acidity of the water will be a variable. In fact, any factor that is different from one group to another is a **variable.** If your hypothesis is correct, then the acidity of the water and plant growth are *dependant variables.* The amount a plant grows is dependent on the acidity of the water. However, the amount of water each plant receives and the amount of sunlight each plant receives are *independent variables.* Either of these factors could change without affecting the other factor.

Sometimes, the nature of an investigation makes a controlled experiment impossible. For example, the Earth's core is surrounded by thousands of meters of rock. Under such circumstances, a hypothesis may be tested by making detailed observations.

4 Analyze the Results

After you have completed your experiments, made your observations, and collected your data, you must analyze all the information you have gathered. Tables and graphs are often used in this step to organize the data.

5 Draw Conclusions

After analyzing your data, you can determine if your results support your hypothesis. If your hypothesis is supported, you (or others) might want to repeat the observations or experiments to verify your results. If your hypothesis is not supported by the data, you may have to check your procedure for errors. You may even have to reject your hypothesis and make a new one. If you cannot draw a conclusion from your results, you may have to try the investigation again or carry out further observations or experiments.

6 Communicate Results

After any scientific investigation, you should report your results. By preparing a written or oral report, you let others know what you have learned. They may repeat your investigation to see if they get the same results. Your report may even lead to another question and then to another investigation.

Scientific Methods in Action

Scientific methods contain loops in which several steps may be repeated over and over again. In some cases, certain steps are unnecessary. Thus, there is not a "straight line" of steps. For example, sometimes scientists find that testing one hypothesis raises new questions and new hypotheses to be tested. And sometimes, testing the hypothesis leads directly to a conclusion. Furthermore, the steps in scientific methods are not always used in the same order. Follow the steps in the diagram, and see how many different directions scientific methods can take you.

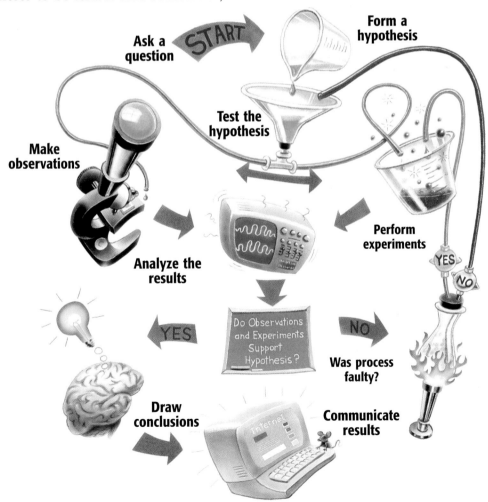

Appendix

Periodic Table of the Elements

Each square on the table includes an element's name, chemical symbol, atomic number, and atomic mass.

The color of the chemical symbol indicates the physical state at room temperature. Carbon is a solid.

6	Atomic number
C	Chemical symbol
Carbon	Element name
12.0	Atomic mass

The background color indicates the type of element. Carbon is a nonmetal.

Background
- Metals
- Metalloids
- Nonmetals

Chemical symbol
- Solid
- Liquid
- Gas

Period 1

·1
H
Hydrogen
1.0

	Group 1	Group 2
Period 2	3 **Li** Lithium 6.9	4 **Be** Beryllium 9.0
Period 3	11 **Na** Sodium 23.0	12 **Mg** Magnesium 24.3

			Group 3	Group 4	Group 5	Group 6	Group 7	Group 8	Group 9
Period 4	19 **K** Potassium 39.1	20 **Ca** Calcium 40.1	21 **Sc** Scandium 45.0	22 **Ti** Titanium 47.9	23 **V** Vanadium 50.9	24 **Cr** Chromium 52.0	25 **Mn** Manganese 54.9	26 **Fe** Iron 55.8	27 **Co** Cobalt 58.9
Period 5	37 **Rb** Rubidium 85.5	38 **Sr** Strontium 87.6	39 **Y** Yttrium 88.9	40 **Zr** Zirconium 91.2	41 **Nb** Niobium 92.9	42 **Mo** Molybdenum 95.9	43 **Tc** Technetium (98)	44 **Ru** Ruthenium 101.1	45 **Rh** Rhodium 102.9
Period 6	55 **Cs** Cesium 132.9	56 **Ba** Barium 137.3	57 **La** Lanthanum 138.9	72 **Hf** Hafnium 178.5	73 **Ta** Tantalum 180.9	74 **W** Tungsten 183.8	75 **Re** Rhenium 186.2	76 **Os** Osmium 190.2	77 **Ir** Iridium 192.2
Period 7	87 **Fr** Francium (223)	88 **Ra** Radium (226)	89 **Ac** Actinium (227)	104 **Rf** Rutherfordium (261)	105 **Db** Dubnium (262)	106 **Sg** Seaborgium (266)	107 **Bh** Bohrium (264)	108 **Hs** Hassium (277)	109 **Mt** Meitnerium (268)

A row of elements is called a *period*.

A column of elements is called a *group* or *family*.

Values in parentheses are the mass numbers of those radioactive elements' most stable or most common isotopes.

These elements are placed below the table to allow the table to be narrower.

Lanthanides

58 **Ce** Cerium 140.1	59 **Pr** Praseodymium 140.9	60 **Nd** Neodymium 144.2	61 **Pm** Promethium (145)	62 **Sm** Samarium 150.4

Actinides

90 **Th** Thorium 232.0	91 **Pa** Protactinium 231.0	92 **U** Uranium 238.0	93 **Np** Neptunium (237)	94 **Pu** Plutonium (244)

Topic: **Periodic Table**
Go To: go.hrw.com
Keyword: **HN0 PERIODIC**
Visit the HRW Web site for
updates on the periodic table.

Group 18

| 2 |
| **He** |
| Helium |
| 4.0 |

This zigzag line
reminds you where
the metals, nonmetals,
and metalloids are.

Group 13	Group 14	Group 15	Group 16	Group 17
5	6	7	8	9
B	**C**	N	O	F
Boron	Carbon	Nitrogen	Oxygen	Fluorine
10.8	12.0	14.0	16.0	19.0

| 10 |
| Ne |
| Neon |
| 20.2 |

13	14	15	16	17	18
Al	**Si**	**P**	**S**	Cl	Ar
Aluminum	Silicon	Phosphorus	Sulfur	Chlorine	Argon
27.0	28.1	31.0	32.1	35.5	39.9

Group 10	Group 11	Group 12					
28	29	30	31	32	33	34	35	36
Ni	**Cu**	**Zn**	**Ga**	**Ge**	**As**	**Se**	**Br**	Kr
Nickel	Copper	Zinc	Gallium	Germanium	Arsenic	Selenium	Bromine	Krypton
58.7	63.5	65.4	69.7	72.6	74.9	79.0	79.9	83.8
46	47	48	49	50	51	52	53	54
Pd	**Ag**	**Cd**	**In**	**Sn**	**Sb**	**Te**	**I**	Xe
Palladium	Silver	Cadmium	Indium	Tin	Antimony	Tellurium	Iodine	Xenon
106.4	107.9	112.4	114.8	118.7	121.8	127.6	126.9	131.3
78	79	80	81	82	83	84	85	86
Pt	**Au**	**Hg**	**Tl**	**Pb**	**Bi**	**Po**	**At**	Rn
Platinum	Gold	Mercury	Thallium	Lead	Bismuth	Polonium	Astatine	Radon
195.1	197.0	200.6	204.4	207.2	209.0	(209)	(210)	(222)
110	111	112	113	114	115			
Ds	**Uuu**	**Uub**	**Uut**	**Uuq**	**Uup**			
Darmstadtium	Unununium	Ununbium	Ununtrium	Ununquadium	Ununpentium			
(281)	(272)	(285)	(284)	(289)	(288)			

The discovery of elements
113, 114, and 115 has been
reported but not confirmed.

The names and three-letter symbols of elements are temporary. They
are based on the atomic numbers of the elements. Official names and
symbols will be approved by an international committee of scientists.

63	64	65	66	67	68	69	70	71
Eu	**Gd**	**Tb**	**Dy**	**Ho**	**Er**	**Tm**	**Yb**	**Lu**
Europium	Gadolinium	Terbium	Dysprosium	Holmium	Erbium	Thulium	Ytterbium	Lutetium
152.0	157.2	158.9	162.5	164.9	167.3	168.9	173.0	175.0
95	96	97	98	99	100	101	102	103
Am	**Cm**	**Bk**	**Cf**	**Es**	**Fm**	**Md**	**No**	**Lr**
Americium	Curium	Berkelium	Californium	Einsteinium	Fermium	Mendelevium	Nobelium	Lawrencium
(243)	(247)	(247)	(251)	(252)	(257)	(258)	(259)	(262)

Appendix

Making Charts and Graphs

Pie Charts

A pie chart shows how each group of data relates to all of the data. Each part of the circle forming the chart represents a category of the data. The entire circle represents all of the data. For example, a biologist studying a hardwood forest in Wisconsin found that there were five different types of trees. The data table at right summarizes the biologist's findings.

Wisconsin Hardwood Trees	
Type of tree	Number found
Oak	600
Maple	750
Beech	300
Birch	1,200
Hickory	150
Total	3,000

How to Make a Pie Chart

1 To make a pie chart of these data, first find the percentage of each type of tree. Divide the number of trees of each type by the total number of trees, and multiply by 100.

$$\frac{600 \text{ oak}}{3,000 \text{ trees}} \times 100 = 20\%$$

$$\frac{750 \text{ maple}}{3,000 \text{ trees}} \times 100 = 25\%$$

$$\frac{300 \text{ beech}}{3,000 \text{ trees}} \times 100 = 10\%$$

$$\frac{1,200 \text{ birch}}{3,000 \text{ trees}} \times 100 = 40\%$$

$$\frac{150 \text{ hickory}}{3,000 \text{ trees}} \times 100 = 5\%$$

2 Now, determine the size of the wedges that make up the pie chart. Multiply each percentage by 360°. Remember that a circle contains 360°.

$20\% \times 360° = 72°$ $25\% \times 360° = 90°$

$10\% \times 360° = 36°$ $40\% \times 360° = 144°$

$5\% \times 360° = 18°$

3 Check that the sum of the percentages is 100 and the sum of the degrees is 360.

$20\% + 25\% + 10\% + 40\% + 5\% = 100\%$

$72° + 90° + 36° + 144° + 18° = 360°$

4 Use a compass to draw a circle and mark the center of the circle.

5 Then, use a protractor to draw angles of 72°, 90°, 36°, 144°, and 18° in the circle.

6 Finally, label each part of the chart, and choose an appropriate title.

A Community of Wisconsin Hardwood Trees

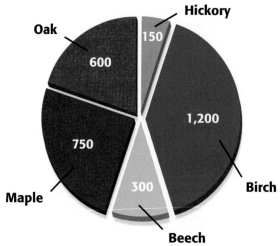

Line Graphs

Line graphs are most often used to demonstrate continuous change. For example, Mr. Smith's students analyzed the population records for their hometown, Appleton, between 1900 and 2000. Examine the data at right.

Because the year and the population change, they are the *variables*. The population is determined by, or dependent on, the year. Therefore, the population is called the **dependent variable,** and the year is called the **independent variable.** Each set of data is called a **data pair.** To prepare a line graph, you must first organize data pairs into a table like the one at right.

Population of Appleton, 1900–2000	
Year	**Population**
1900	1,800
1920	2,500
1940	3,200
1960	3,900
1980	4,600
2000	5,300

How to Make a Line Graph

1 Place the independent variable along the horizontal (*x*) axis. Place the dependent variable along the vertical (*y*) axis.

2 Label the *x*-axis "Year" and the *y*-axis "Population." Look at your largest and smallest values for the population. For the *y*-axis, determine a scale that will provide enough space to show these values. You must use the same scale for the entire length of the axis. Next, find an appropriate scale for the *x*-axis.

3 Choose reasonable starting points for each axis.

4 Plot the data pairs as accurately as possible.

5 Choose a title that accurately represents the data.

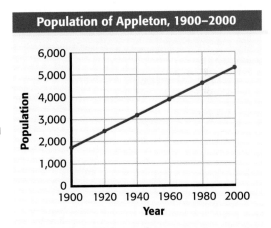

How to Determine Slope

Slope is the ratio of the change in the *y*-value to the change in the *x*-value, or "rise over run."

1 Choose two points on the line graph. For example, the population of Appleton in 2000 was 5,300 people. Therefore, you can define point *a* as (2000, 5,300). In 1900, the population was 1,800 people. You can define point *b* as (1900, 1,800).

2 Find the change in the *y*-value. (*y* at point *a*) − (*y* at point *b*) = 5,300 people − 1,800 people = 3,500 people

3 Find the change in the *x*-value. (*x* at point *a*) − (*x* at point *b*) = 2000 − 1900 = 100 years

4 Calculate the slope of the graph by dividing the change in *y* by the change in *x*.

$$slope = \frac{change\ in\ y}{change\ in\ x}$$

$$slope = \frac{3,500\ people}{100\ years}$$

$$slope = 35\ people\ per\ year$$

In this example, the population in Appleton increased by a fixed amount each year. The graph of these data is a straight line. Therefore, the relationship is **linear.** When the graph of a set of data is not a straight line, the relationship is **nonlinear.**

Using Algebra to Determine Slope

The equation in step 4 may also be arranged to be

$$y = kx$$

where y represents the change in the y-value, k represents the slope, and x represents the change in the x-value.

$$slope = \frac{change\ in\ y}{change\ in\ x}$$

$$k = \frac{y}{x}$$

$$k \times x = \frac{y \times x}{x}$$

$$kx = y$$

Bar Graphs

Bar graphs are used to demonstrate change that is not continuous. These graphs can be used to indicate trends when the data cover a long period of time. A meteorologist gathered the precipitation data shown here for Hartford, Connecticut, for April 1–15, 1996, and used a bar graph to represent the data.

Precipitation in Hartford, Connecticut April 1–15, 1996			
Date	Precipitation (cm)	Date	Precipitation (cm)
April 1	0.5	April 9	0.25
April 2	1.25	April 10	0.0
April 3	0.0	April 11	1.0
April 4	0.0	April 12	0.0
April 5	0.0	April 13	0.25
April 6	0.0	April 14	0.0
April 7	0.0	April 15	6.50
April 8	1.75		

How to Make a Bar Graph

1 Use an appropriate scale and a reasonable starting point for each axis.

2 Label the axes, and plot the data.

3 Choose a title that accurately represents the data.

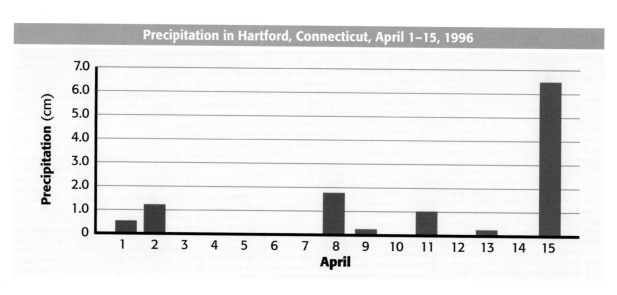

Appendix

Math Refresher

Science requires an understanding of many math concepts. The following pages will help you review some important math skills.

Averages

An **average,** or **mean,** simplifies a set of numbers into a single number that *approximates* the value of the set.

Example: Find the average of the following set of numbers: 5, 4, 7, and 8.

Step 1: Find the sum.
$$5 + 4 + 7 + 8 = 24$$

Step 2: Divide the sum by the number of numbers in your set. Because there are four numbers in this example, divide the sum by 4.
$$\frac{24}{4} = 6$$

The average, or mean, is **6.**

Ratios

A **ratio** is a comparison between numbers, and it is usually written as a fraction.

Example: Find the ratio of thermometers to students if you have 36 thermometers and 48 students in your class.

Step 1: Make the ratio.
$$\frac{36 \text{ thermometers}}{48 \text{ students}}$$

Step 2: Reduce the fraction to its simplest form.
$$\frac{36}{48} = \frac{36 \div 12}{48 \div 12} = \frac{3}{4}$$

The ratio of thermometers to students is **3 to 4,** or $\frac{3}{4}$. The ratio may also be written in the form 3:4.

Proportions

A **proportion** is an equation that states that two ratios are equal.
$$\frac{3}{1} = \frac{12}{4}$$

To solve a proportion, first multiply across the equal sign. This is called *cross-multiplication.* If you know three of the quantities in a proportion, you can use cross-multiplication to find the fourth.

Example: Imagine that you are making a scale model of the solar system for your science project. The diameter of Jupiter is 11.2 times the diameter of the Earth. If you are using a plastic-foam ball that has a diameter of 2 cm to represent the Earth, what must the diameter of the ball representing Jupiter be?
$$\frac{11.2}{1} = \frac{x}{2 \text{ cm}}$$

Step 1: Cross-multiply.
$$\frac{11.2}{1} \diagdown \!\!\!\!\!\diagup \frac{x}{2}$$
$$11.2 \times 2 = x \times 1$$

Step 2: Multiply.
$$22.4 = x \times 1$$

Step 3: Isolate the variable by dividing both sides by 1.
$$x = \frac{22.4}{1}$$
$$x = 22.4 \text{ cm}$$

You will need to use a ball that has a diameter of **22.4** cm to represent Jupiter.

Percentages

A **percentage** is a ratio of a given number to 100.

    Example: What is 85% of 40?

Step 1: Rewrite the percentage by moving the decimal point two places to the left.

    0.85

Step 2: Multiply the decimal by the number that you are calculating the percentage of.

    $0.85 \times 40 = 34$

85% of 40 is **34.**

Decimals

To **add** or **subtract decimals,** line up the digits vertically so that the decimal points line up. Then, add or subtract the columns from right to left. Carry or borrow numbers as necessary.

    Example: Add the following numbers: 3.1415 and 2.96.

Step 1: Line up the digits vertically so that the decimal points line up.

$$\begin{array}{r} 3.1415 \\ + 2.96 \\ \hline \end{array}$$

Step 2: Add the columns from right to left, and carry when necessary.

$$\begin{array}{r} {\scriptstyle 1\ \ 1} \\ 3.1415 \\ + 2.96 \\ \hline 6.1015 \end{array}$$

The sum is **6.1015.**

Fractions

Numbers tell you how many; **fractions** tell you *how much of a whole.*

    Example: Your class has 24 plants. Your teacher instructs you to put 5 plants in a shady spot. What fraction of the plants in your class will you put in a shady spot?

Step 1: In the denominator, write the total number of parts in the whole.

$$\frac{?}{24}$$

Step 2: In the numerator, write the number of parts of the whole that are being considered.

$$\frac{5}{24}$$

So, $\frac{5}{24}$ of the plants will be in the shade.

Reducing Fractions

It is usually best to express a fraction in its simplest form. Expressing a fraction in its simplest form is called *reducing* a fraction.

    Example: Reduce the fraction $\frac{30}{45}$ to its simplest form.

Step 1: Find the largest whole number that will divide evenly into both the numerator and denominator. This number is called the *greatest common factor* (GCF).

Factors of the numerator 30:
    1, 2, 3, 5, 6, 10, **15,** 30

Factors of the denominator 45:
    1, 3, 5, 9, **15,** 45

Step 2: Divide both the numerator and the denominator by the GCF, which in this case is 15.

$$\frac{30}{45} = \frac{30 \div 15}{45 \div 15} = \frac{2}{3}$$

Thus, $\frac{30}{45}$ reduced to its simplest form is $\frac{2}{3}$.

Appendix

Adding and Subtracting Fractions

To **add** or **subtract fractions** that have the **same denominator,** simply add or subtract the numerators.

Examples:

$$\frac{3}{5} + \frac{1}{5} = ? \text{ and } \frac{3}{4} - \frac{1}{4} = ?$$

Step 1: Add or subtract the numerators.

$$\frac{3}{5} + \frac{1}{5} = \frac{4}{?} \text{ and } \frac{3}{4} - \frac{1}{4} = \frac{2}{?}$$

Step 2: Write the sum or difference over the denominator.

$$\frac{3}{5} + \frac{1}{5} = \frac{4}{5} \text{ and } \frac{3}{4} - \frac{1}{4} = \frac{2}{4}$$

Step 3: If necessary, reduce the fraction to its simplest form.

$\frac{4}{5}$ cannot be reduced, and $\frac{2}{4} = \frac{1}{2}$.

To **add** or **subtract fractions** that have **different denominators,** first find the least common denominator (LCD).

Examples:

$$\frac{1}{2} + \frac{1}{6} = ? \text{ and } \frac{3}{4} - \frac{2}{3} = ?$$

Step 1: Write the equivalent fractions that have a common denominator.

$$\frac{3}{6} + \frac{1}{6} = ? \text{ and } \frac{9}{12} - \frac{8}{12} = ?$$

Step 2: Add or subtract the fractions.

$$\frac{3}{6} + \frac{1}{6} = \frac{4}{6} \text{ and } \frac{9}{12} - \frac{8}{12} = \frac{1}{12}$$

Step 3: If necessary, reduce the fraction to its simplest form.

The fraction $\frac{4}{6} = \frac{2}{3}$, and $\frac{1}{12}$ cannot be reduced.

Multiplying Fractions

To **multiply fractions,** multiply the numerators and the denominators together, and then reduce the fraction to its simplest form.

Example:

$$\frac{5}{9} \times \frac{7}{10} = ?$$

Step 1: Multiply the numerators and denominators.

$$\frac{5}{9} \times \frac{7}{10} = \frac{5 \times 7}{9 \times 10} = \frac{35}{90}$$

Step 2: Reduce the fraction.

$$\frac{35}{90} = \frac{35 \div 5}{90 \div 5} = \frac{7}{18}$$

Dividing Fractions

To **divide fractions,** first rewrite the divisor (the number you divide by) upside down. This number is called the *reciprocal* of the divisor. Then multiply and reduce if necessary.

Example:

$$\frac{5}{8} \div \frac{3}{2} = ?$$

Step 1: Rewrite the divisor as its reciprocal.

$$\frac{3}{2} \rightarrow \frac{2}{3}$$

Step 2: Multiply the fractions.

$$\frac{5}{8} \times \frac{2}{3} = \frac{5 \times 2}{8 \times 3} = \frac{10}{24}$$

Step 3: Reduce the fraction.

$$\frac{10}{24} = \frac{10 \div 2}{24 \div 2} = \frac{5}{12}$$

Scientific Notation

Scientific notation is a short way of representing very large and very small numbers without writing all of the place-holding zeros.

> **Example:** Write 653,000,000 in scientific notation.

Step 1: Write the number without the place-holding zeros.

$$653$$

Step 2: Place the decimal point after the first digit.

$$6.53$$

Step 3: Find the exponent by counting the number of places that you moved the decimal point.

$$6.53000000$$

The decimal point was moved eight places to the left. Therefore, the exponent of 10 is positive 8. If you had moved the decimal point to the right, the exponent would be negative.

Step 4: Write the number in scientific notation.

$$\textbf{6.53} \times \textbf{10}^\textbf{8}$$

Area

Area is the number of square units needed to cover the surface of an object.

Formulas:

area of a square = side × side
area of a rectangle = length × width
area of a triangle = $\frac{1}{2}$ × base × height

Examples: Find the areas.

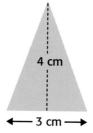

Triangle

$area = \frac{1}{2}$ × *base × height*

$area = \frac{1}{2}$ × 3 cm × 4 cm

$area =$ **6 cm²**

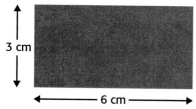

Rectangle
area = length × width
area = 6 cm × 3 cm
area = **18 cm²**

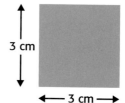

Square
area = side × side
area = 3 cm × 3 cm
area = **9 cm²**

Volume

Volume is the amount of space that something occupies.

Formulas:

volume of a cube = side × side × side

volume of a prism = area of base × height

Examples:

Find the volume of the solids.

Cube
volume = side × side × side
volume = 4 cm × 4 cm × 4 cm
volume = **64 cm³**

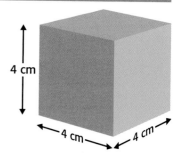

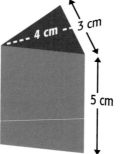

Prism
volume = area of base × height
volume = (area of triangle) × height
volume = ($\frac{1}{2}$ × 3 cm × 4 cm) × 5 cm
volume = 6 cm² × 5 cm
volume = **30 cm³**

Physical Science Laws and Principles

Law of Conservation of Mass

Mass cannot be created or destroyed during ordinary chemical or physical changes.

The total mass in a closed system is always the same no matter how many physical changes or chemical reactions occur.

Law of Conservation of Energy

Energy can be neither created nor destroyed.

The total amount of energy in a closed system is always the same. Energy can be changed from one form to another, but all of the different forms of energy in a system always add up to the same total amount of energy no matter how many energy conversions occur.

Law of Universal Gravitation

All objects in the universe attract each other by a force called *gravity*. The size of the force depends on the masses of the objects and the distance between the objects.

The first part of the law explains why lifting a bowling ball is much harder than lifting a marble. Because the bowling ball has a much larger mass than the marble does, the amount of gravity between the Earth and the bowling ball is greater than the amount of gravity between the Earth and the marble.

The second part of the law explains why a satellite can remain in orbit around the Earth. The satellite is carefully placed at a distance great enough to prevent the Earth's gravity from immediately pulling the satellite down but small enough to prevent the satellite from completely escaping the Earth's gravity and wandering off into space.

Newton's Laws of Motion

Newton's first law of motion states that an object at rest remains at rest and an object in motion remains in motion at constant speed and in a straight line unless acted on by an unbalanced force.

The first part of the law explains why a football will remain on a tee until it is kicked off or until a gust of wind blows it off.

The second part of the law explains why a bike rider will continue moving forward after the bike comes to an abrupt stop. Gravity and the friction of the sidewalk will eventually stop the rider.

Newton's second law of motion states that the acceleration of an object depends on the mass of the object and the amount of force applied.

The first part of the law explains why the acceleration of a 4 kg bowling ball will be greater than the acceleration of a 6 kg bowling ball if the same force is applied to both balls.

The second part of the law explains why the acceleration of a bowling ball will be larger if a larger force is applied to the bowling ball.

The relationship of acceleration (a) to mass (m) and force (F) can be expressed mathematically by the following equation:

$$acceleration = \frac{force}{mass}, \text{ or } a = \frac{F}{m}$$

This equation is often rearranged to the form

$$force = mass \times acceleration, \text{ or } F = m \times a$$

Newton's third law of motion states that whenever one object exerts a force on a second object, the second object exerts an equal and opposite force on the first.

This law explains that a runner is able to move forward because of the equal and opposite force that the ground exerts on the runner's foot after each step.

Law of Reflection

The law of reflection states that the angle of incidence is equal to the angle of reflection. This law explains why light reflects off a surface at the same angle that the light strikes the surface.

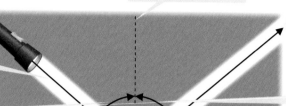

The beam of light traveling toward the mirror is called the *incident beam.*

A line perpendicular to the mirror's surface is called the *normal.*

The beam of light reflected off the mirror is called the *reflected beam.*

The angle between the incident beam and the normal is called the *angle of incidence.*

The angle between the reflected beam and the normal is called the *angle of reflection.*

Charles's Law

Charles's law states that for a fixed amount of gas at a constant pressure, the volume of the gas increases as the temperature of the gas increases. Likewise, the volume of the gas decreases as the temperature of the gas decreases.

If a basketball that was inflated indoors is left outside on a cold winter day, the air particles inside the ball will move more slowly. They will hit the sides of the basketball less often and with less force. The ball will get smaller as the volume of the air decreases.

Boyle's Law

Boyle's law states that for a fixed amount of gas at a constant temperature, the volume of a gas increases as the pressure of the gas decreases. Likewise, the volume of a gas decreases as its pressure increases.

If an inflated balloon is pulled down to the bottom of a swimming pool, the pressure of the water on the balloon increases. The pressure of the air particles inside the balloon must increase to match that of the water outside, so the volume of the air inside the balloon decreases.

Pascal's Principle

Pascal's principle states that a change in pressure at any point in an enclosed fluid will be transmitted equally to all parts of that fluid.

When a mechanic uses a hydraulic jack to raise an automobile off the ground, he or she increases the pressure on the fluid in the jack by pushing on the jack handle. The pressure is transmitted equally to all parts of the fluid-filled jacking system. As fluid presses the jack plate against the frame of the car, the car is lifed off the ground.

Archimedes' Principle

Archimedes' principle states that the buoyant force on an object in a fluid is equal to the weight of the volume of fluid that the object displaces.

A person floating in a swimming pool displaces 20 L of water. The weight of that volume of water is about 200 N. Therefore, the buoyant force on the person is 200 N.

Bernoulli's Principle

Bernoulli's principle states that as the speed of a moving fluid increases, the fluid's pressure decreases.

The lift on an airplane wing or on a Frisbee® can be explained in part by using Bernoulli's principle. Because of the shape of the Frisbee, the air moving over the top of the Frisbee must travel farther than the air below the Frisbee in the same amount of time. In other words, the air above the Frisbee is moving faster than the air below it. This faster-moving air above the Frisbee exerts less pressure than the slower-moving air below it does. The resulting increased pressure below exerts an upward force and pushes the Frisbee up.

Useful Equations

Average speed

$$average\ speed = \frac{total\ distance}{total\ time}$$

Example: A bicycle messenger traveled a distance of 136 km in 8 h. What was the messenger's average speed?

$$\frac{136\ km}{8\ h} = 17\ km/h$$

The messenger's average speed was **17 km/h.**

Average acceleration

$$\frac{average}{acceleration} = \frac{final\ velocity - starting\ velocity}{time\ it\ takes\ to\ change\ velocity}$$

Example: Calculate the average acceleration of an Olympic 100 m dash sprinter who reaches a velocity of 20 m/s south at the finish line. The race was in a straight line and lasted 10 s.

$$\frac{20\ m/s - 0\ m/s}{10s} = 2\ m/s/s$$

The sprinter's average acceleration is **2 m/s/s south.**

Net force

Forces in the Same Direction
When forces are in the same direction, add the forces together to determine the net force.

Example: Calculate the net force on a stalled car that is being pushed by two people. One person is pushing with a force of 13 N northwest, and the other person is pushing with a force of 8 N in the same direction.

$$13\ N + 8\ N = 21\ N$$

The net force is **21 N northwest.**

Forces in Opposite Directions
When forces are in opposite directions, subtract the smaller force from the larger force to determine the net force. The net force will be in the direction of the larger force.

Example: Calculate the net force on a rope that is being pulled on each end. One person is pulling on one end of the rope with a force of 12 N south. Another person is pulling on the opposite end of the rope with a force of 7 N north.

$$12\ N - 7\ N = 5\ N$$

The net force is **5 N south.**

Work

Work is done by exerting a force through a distance. Work has units of joules (J), which are equivalent to Newton-meters.

$$Work = F \times d$$

Example: Calculate the amount of work done by a man who lifts a 100 N toddler 1.5 m off the floor.

$Work = 100 \text{ N} \times 1.5 \text{ m} = 150 \text{ N•m} = 150 \text{ J}$

The man did **150 J** of work.

Power

Power is the rate at which work is done. Power is measured in watts (W), which are equivalent to joules per second.

$$P = \frac{Work}{t}$$

Example: Calculate the power of a weight-lifter who raises a 300 N barbell 2.1 m off the floor in 1.25 s.

$Work = 300 \text{ N} \times 2.1 \text{ m} = 630 \text{ N•m} = 630 \text{ J}$

$$P = \frac{630 \text{ J}}{1.25 \text{ s}} = \frac{504 \text{ J}}{\text{s}} = 504 \text{ W}$$

The weightlifter has **504 W** of power.

Pressure

Pressure is the force exerted over a given area. The SI unit for pressure is the pascal (Pa).

$$pressure = \frac{force}{area}$$

Example: Calculate the pressure of the air in a soccer ball if the air exerts a force of 25,000 N over an area of 0.15 m².

$$pressure = \frac{25,000 \text{ N}}{0.15 \text{ m}^2} = \frac{167,000 \text{ N}}{\text{m}^2} = 167,000 \text{ Pa}$$

The pressure of the air inside the soccer ball is **167,000 Pa.**

Density

$$density = \frac{mass}{volume}$$

Example: Calculate the density of a sponge that has a mass of 10 g and a volume of 40 cm³.

$$\frac{10 \text{ g}}{40 \text{ cm}^3} = \frac{0.25 \text{ g}}{\text{cm}^3}$$

The density of the sponge is $\frac{0.25 \text{ g}}{\text{cm}^3}$.

Concentration

$$concentration = \frac{mass \text{ of } solute}{volume \text{ of } solvent}$$

Example: Calculate the concentration of a solution in which 10 g of sugar is dissolved in 125 mL of water.

$$\frac{10 \text{ g of sugar}}{125 \text{ mL of water}} = \frac{0.08 \text{ g}}{\text{mL}}$$

The concentration of this solution is $\frac{0.08 \text{ g}}{\text{mL}}$.

Glossary

Glossary

A

absorption in optics, the transfer of light energy to particles of matter (76)

amplitude the maximum distance that the particles of a wave's medium vibrate from their rest position (10)

C

concave lens a lens that is thinner in the middle than at the edges (103)

concave mirror a mirror that is curved inward like the inside of a spoon (100)

convex lens a lens that is thicker in the middle than at the edges (102)

convex mirror a mirror that is curved outward like the back of a spoon (101)

D

decibel the most common unit used to measure loudness (symbol, dB) (40)

diffraction a change in the direction of a wave when the wave finds an obstacle or an edge, such as an opening (16, 79)

Doppler effect an observed change in the frequency of a wave when the source or observer is moving (38)

E

echo a reflected sound wave (42)

echolocation the process of using reflected sound waves to find objects; used by animals such as bats (43)

electromagnetic spectrum all of the frequencies or wavelengths of electromagnetic radiation (66)

electromagnetic wave a wave that consists of electric and magnetic fields that vibrate at right angles to each other (62)

F

farsightedness a condition in which the lens of the eye focuses distant objects behind rather than on the retina (105)

frequency the number of waves produced in a given amount of time (12)

H

hologram a piece of film that produces a three-dimensional image of an object; made by using laser light (111)

I

interference the combination of two or more waves that results in a single wave (17, 44, 80)

L

laser a device that produces intense light of only one wavelength and color (109)

lens a transparent object that refracts light waves such that they converge or diverge to create an image (102)

longitudinal wave a wave in which the particles of the medium vibrate parallel to the direction of wave motion (8)

loudness the extent to which a sound can be heard (39)

M

medium a physical environment in which phenomena occur (5, 32)

N

nearsightedness a condition in which the lens of the eye focuses distant objects in front of rather than on the retina (105)

noise a sound that consists of a random mix of frequencies (51)

O

opaque (oh PAYK) describes an object that is not transparent or translucent (83)

P

pigment a substance that gives another substance or a mixture its color (86)

pitch a measure of how high or low a sound is perceived to be, depending on the frequency of the sound wave (37)

plane mirror a mirror that has a flat surface (99)

R

radiation the transfer of energy as electromagnetic waves (63)

reflection the bouncing back of a ray of light, sound, or heat when the ray hits a surface that it does not go through (14, 74)

refraction the bending of a wave as the wave passes between two substances in which the speed of the wave differs (15, 77)

resonance a phenomenon that occurs when two objects naturally vibrate at the same frequency; the sound produced by one object causes the other object to vibrate (19, 46)

S

scattering an interaction of light with matter that causes light to change its energy, direction of motion, or both (76)

sonic boom the explosive sound heard when a shock wave from an object traveling faster than the speed of sound reaches a person's ears (45)

sound quality the result of the blending of several pitches through interference (48)

sound wave a longitudinal wave that is caused by vibrations and that travels through a material medium (31)

standing wave a pattern of vibration that simulates a wave that is standing still (18, 46)

T

translucent (trans LOO suhnt) describes matter that transmits light but that does not transmit an image (83)

transmission the passing of light or other form of energy through matter (82)

transparent describes matter that allows light to pass through with little interference (83)

transverse wave a wave in which the particles of the medium move perpendicularly to the direction the wave is traveling (7)

W

wave a periodic disturbance in a solid, liquid, or gas as energy is transmitted through a medium (4)

wavelength the distance from any point on a wave to an identical point on the next wave (11)

wave speed the speed at which a wave travels through a medium (12)

Glossary

Spanish Glossary

A

absorption/absorción en la óptica, la transferencia de energía luminosa a las partículas de materia (76)

amplitude/amplitud la distancia máxima a la que vibran las partículas del medio de una onda a partir de su posición de reposo (10)

C

concave lens/lente cóncava una lente que es más delgada en la parte media que en los bordes (103)

concave mirror/espejo cóncavo un espejo que está curvado hacia adentro como la parte interior de una cuchara (100)

convex lens/lente convexa una lente que es más gruesa en la parte media que en los bordes (102)

convex mirror/espejo convexo un espejo que está curvado hacia fuera como la parte de atrás de una cuchara (101)

D

decibel/decibel la unidad más común que se usa para medir el volumen del sonido (símbolo: dB) (40)

diffraction/difracción un cambio en la dirección de una onda cuando ésta se encuentra con un obstáculo o un borde, tal como una abertura (16, 79)

Doppler effect/efecto Doppler un cambio que se observa en la frecuencia de una onda cuando la fuente o el observador está en movimiento (38)

E

echo/eco una onda de sonido reflejada (42)

echolocation/ecolocación el proceso de usar ondas de sonido reflejadas para buscar objetos; utilizado por animales tales como los murciélagos (43)

electromagnetic spectrum/espectro electromagnético todas las frecuencias o longitudes de onda de la radiación electromagnética (66)

electromagnetic wave/onda electromagnética una onda que está formada por campos eléctricos y magnéticos que vibran formando un ángulo recto unos con otros (62)

F

farsightedness/hipermetropía condición en la que el cristalino del ojo enfoca los objetos lejanos detrás de la retina en lugar de en ella (105)

frequency/frecuencia el número de ondas producidas en una cantidad de tiempo determinada (12)

H

hologram/holograma una porción de película que produce una imagen tridimensional de un objeto mediante luz láser (111)

I

interference/interferencia la combinación de dos o más ondas que resulta en una sola onda (17, 44, 80)

L

laser/láser un aparato que produce una luz intensa de únicamente una longitud de onda y color (109)

lens/lente un objeto transparente que refracta las ondas de luz de modo que converjan o diverjan para crear una imagen (102)

longitudinal wave/onda longitudinal una onda en la que las partículas del medio vibran paralelamente a la dirección del movimiento de la onda (8)

loudness/volumen el grado al que se escucha un sonido (39)

M

medium/medio un ambiente físico en el que ocurren fenómenos (5, 32)

N

nearsightedness/miopía condición en la que el cristalino del ojo enfoca los objetos lejanos delante de la retina en lugar de en ella (105)

noise/ruido un sonido que está constituido por una mezcla aleatoria de frecuencias (51)

O

opaque/opaco término que describe un objeto que no es transparente ni translúcido (83)

P

pigment/pigmento una substancia que le da color a otra substancia o mezcla (86)

pitch/altura tonal una medida de qué tan agudo o grave se percibe un sonido, dependiendo de la frecuencia de la onda sonora (37)

plane mirror/espejo plano un espejo que tiene una superficie plana (99)

R

radiation/radiación la transferencia de energía en forma de ondas electromagnéticas (63)

reflection/reflexión el rebote de un rayo de luz, sonido o calor cuando el rayo golpea una superficie pero no la atraviesa (14, 74)

refraction/refracción el curvamiento de una onda cuando ésta pasa entre dos substancias en las que su velocidad difiere (15, 77)

resonance/resonancia un fenómeno que ocurre cuando dos objetos vibran naturalmente a la misma frecuencia; el sonido producido por un objeto hace que el otro objeto vibre (19, 46)

S

scattering/dispersión una interacción de la luz con la materia que hace que la luz cambie su energía, la dirección del movimiento o ambas (76)

sonic boom/estampido sónico el sonido explosivo que se escucha cuando la onda de choque de un objeto que se desplaza a una velocidad superior a la de la luz llega a los oídos de una persona (45)

sound quality/calidad del sonido el resultado de la combinación de varios tonos por medio de la interferencia (48)

sound wave/onda de sonido una onda longitudinal que se origina debido a vibraciones y que se desplaza a través de un medio material (31)

standing wave/onda estacionaria un patrón de vibración que simula una onda que está parada (18, 46)

T

translucent/traslúcido término que describe la materia que transmite luz, pero que no transmite una imagen (83)

transmission/transmisión el paso de la luz u otra forma de energía a través de la materia (82)

transparent/transparente término que describe materia que permite el paso de la luz con poca interferencia (83)

transverse wave/onda transversal una onda en la que las partículas del medio se mueven perpendicularmente respecto a la dirección en la que se desplaza la onda (7)

W

wave/onda una perturbación periódica en un sólido, líquido o gas que se transmite a través de un medio en forma de energía (4)

wavelength/longitud de onda la distancia entre cualquier punto de una onda y un punto idéntico en la onda siguiente (11)

wave speed/rapidez de onda la rapidez a la cual viaja una onda a través de un medio (12)

Index

Index

energy
law of conservation of, 157
from the sun, 65, **65**
wave, 4–6, **4, 5, 6,** 10–12
English units, **143**
enzymes, 94
experimental groups, 147
experiments, controlled, 147
eyes, 104, **104**
bionic, 122
color deficiency, 106, **106**
surgical correction, 106, **106**
vision problems, 105, **105**

F

Faber, Sandra, 123
Fahrenheit scale, **144**
families (groups), **148–149**
farsightedness, 105, **105**
fields, magnetic, 62–63, **62, 63**
film, 108, **108**
fireflies, 94
fish finders, **43**
flashlights, 100
fluids, Pascal's principle in, 158
FM radio waves, 67
focal length, **100**
focal point, 100, **100, 101**
FoldNote instructions, 137–140,
137, 138, 139, 140
food, inspection of, 94
forces, net, 159
fractions, 154–155
freezing points, **144**
frequencies, 12, **12**
electromagnetic spectrum, 66,
66–67
energy and, 12, **12**
hearing and, 37, **37,** 52–53
on oscilloscopes, 40–41, **40, 41**
resonant, 19, 46, **46**
of sound, 37, **37, 40**
wavelength and, 13
frequency modulation, 67–68
fundamental frequencies, **46,** 48

G

gamma rays, **67,** 72
gamma ray spectrometers, **72**
GCF (greatest common factor), 154
genetics, **106**
glass, 82–84, **82**
Global Positioning System (GPS),
114, **114**
graduated cylinders, 145
grams, **143**

Graphic Organizer instructions,
141–142, **141, 142**
graphs
bar, 152, **152**
circle, 150, **150**
line, 151–152, **151**
slopes of, 151–152
gravitation, law of universal, 157
greatest common factor (GCF), 154
groups, **148–149**
guitars
pickups on, **49**
resonance of, 46–47, **46, 47**
sound quality of, 49, **49**

H

hammers (part of ear), **33**
headlights, car, 100
hearing
deafness and loss of, 34
detecting sounds, 32
ears, **33**
frequency and, 37, **37,** 52–53
protecting, 35, **35**
heat, as infrared radiation, 66, **66,
69**
hectares, **143**
helium-neon lasers, **110**
hertz (Hz), 12, 37
holograms, 111, **111**
Hubble Space Telescope, 123
hypotheses, 146–147

I

illuminated objects, 75, **75**
images
real, 100, **101,** 102, **102, 104**
three-dimensional, 111, **111**
virtual, 99–103, **101, 102, 103**
incidence, 74, **74**
incident beams, **74**
independent variables, 151, **151**
inertia, 157
infrared binoculars, 69
infrared radiation, **66,** 69, **69**
inner ears, **33**
interference, 17, **17,** 44, **44**
constructive, 17, **17, 44,** 80, **80**
destructive, 18, **18, 44**
in light waves, 80, **80**
resonance and, 19, 46–47, **46,
47**
sound barrier and, 45, **45**
in sound waves, 44–46, **44, 45**
in standing waves, 18–19, **18,
19,** 46, **46**
International System of Units (SI),
143

J

jets, 45, **45**

K

kelvins, 144
Kelvin scale, **144**
kilograms (kg), **143**

L

lasers, 109–111, **109, 110, 111**
law of conservation of energy, 157
law of reflection, 74, **74,** 158
law of universal gravitation, 157
laws, scientific, 157–159
laws of motion, Newton's, 157
least common denominator (LCD),
155
length, **143,** 145, **145**
lenses, 98–103, **102**
in cameras, **108**
concave, **102,** 103, **103, 105**
convex, 102, **102, 105,** 116–117
in the eye, 104, **104**
in microscopes, 109, **109**
in telescopes, **109**
light, 62–87, 98–115
absorption and scattering, 76,
76, 77, 82
color addition, 85, **85,** 88–89
colors of objects, 83–84, **84**
color subtraction, 86–87, **86, 87,**
88–89
color television, 85
communication technology, 113,
113
diffraction, 16, **16,** 79, **79**
Einstein on, 95
electromagnetic spectrum, 66,
66–67
as electromagnetic wave, 6, **6,**
62–63, **62, 63**
gamma rays, **67,** 72
infrared radiation, **66,** 69, **69**
interaction with matter, 82–84,
82, 83, 84
interference, 17–19, **17, 18,** 80,
80 (*see also* interference)
labs on, **79, 86,** 88–89,
116–117
from lasers, 109–111, **109, 110,
111**
lenses, 102–103, **102, 103**
luminous *vs.* illuminated objects,
75, **75**

Index

Index

vitamin D, 71
VLA (Very Large Array), 26
vocal sounds, **32**
volume
    of cubes, 156
    formulas for, 156
    of a gas, 158
    of liquids, 145
    units of, **143**

W

water
    freezing point of, **144**
    light refraction in, 77
    refraction of light in, 77–78, **78**
    speed of sound in, **36**
wave equation, 12–13
wavelength, 11, **11**
    diffraction and, 16, **16**
    electromagnetic spectrum, 66,
      66–67
    energy and, 11
    frequency and, 13
    measuring, 11, **11**
waves, 4–19, **4**. *See also*
      wavelength
    absorption and scattering, 42,
      76, **76, 77, 82**
    amplitude of, 10, **10**, 39, **39**
    diffraction of, 16, **16**, 79, **79**
    electromagnetic, 6, **6**
    frequency of, 12–13, **12** (*see*
      also frequencies)
    infrared, **66,** 69, **69**
    interference, 17–19, **17, 18,** 80,
      80
    labs on, **11,** 20–21
    longitudinal, 8, **8, 11, 30,** 31
    mechanical, 5, **5**
    microwaves, **66,** 68–69, **68, 69,**
      94
    radio, **66,** 67–68, **113**
    reflection of, 14, **15,** 74–75, **74,**
      75
    refraction of, 15, **15,** 77–78, **77,**
      78
    resonance, 19
    sound, 5, 8, **8,** 31, **31** (*see also*
      sound waves)
    standing, 18, **18, 19,** 46–47, **46**
    surface, 7, 9, **9**
    transverse, 7, **7, 8,** 11
    ultraviolet, **67,** 71, **71**

visible light, 66, **67,** 70, **70, 71**
wave energy, 4–6, **4, 5, 6,** 10–12
wave speed, 12–13
    work and, 5
    X rays, **67,** 72, **72**
wave speed, 12–13, **12**
white light, 70, 84, **84**
wind instruments, 50, **50, 51**
work, 5, 160

X

X chromosomes, **106**
X rays, **67,** 72, **72**
X-ray technologists, 27

Y

Yeager, Chuck, 36
Yolen, Jane, 58
Young, Thomas, 26

Z

Zavala, Estela, 27

Credits

Abbreviations used: (t) top, (c) center, (b) bottom, (l) left, (r) right, (bkgd) background

PHOTOGRAPHY

Front Cover James Noble/Corbis; (bkgd), Mehau Kulyk/Science Photo Library/Photo Researchers

Skills Practice Lab Teens Sam Dudgeon/HRW

Connection to Astronomy Corbis Images; **Connection to Biology** David M. Phillips/Visuals Unlimited; **Connection to Chemistry** Digital Image copyright © 2005 PhotoDisc; **Connection to Environment** Digital Image copyright © 2005 PhotoDisc; **Connection to Geology** Letraset Phototone; **Connection to Language Arts** Digital Image copyright © 2005 PhotoDisc; **Connection to Meteorology** Digital Image copyright © 2005 PhotoDisc; **Connection to Oceanography** © ICONOTEC; **Connection to Physics** Digital Image copyright © 2005 PhotoDisc

Table of Contents iv (tl), © Jason Childs/Getty Images; iv (bc), John Langford/HRW; v (tr), ©Digital Vision Ltd.; v (b), ©Cameron Davidson/Getty Images; v–vii, Victoria Smith/HRW; x (bl), Sam Dudgeon/HRW; xi (tl), John Langford/HRW; xi (b), Sam Dudgeon/HRW; xii (tl), Victoria Smith/HRW; xii (bl), Stephanie Morris/HRW; xii (br), Sam Dudgeon/HRW; xiii (tl), Patti Murray/Animals, Animals; xiii (tr), Jana Birchum/HRW; xiii (b), Peter Van Steen/HRW

Chapter One 2–3 (all), © Jason Childs/Getty Images; 5 (tr), Robert Mathena/Fundamental Photographs, New York; 5 (bl), © Albert Copley/Visuals Unlimited; 6 (t), NASA; 13 (tr), © Steve Kaufman/CORBIS; 14 (br), Erich Schrempp/Photo Researchers, Inc.; 15 (tl), Richard Megna/Fundamental Photographs; 16 (tc), Educational Development Center; 18 (tl), Richard Megna/Fundamental Photographs; 18 (bl), John Langford/HRW; 20 (br), James H. Karales/Peter Arnold, Inc.; 21 (b), Sam Dudgeon/HRW; 22 (bl), Richard Megna/Fundamental Photographs; 23 (bl), Martin Bough/Fundamental Photographs; 26 (tl), Pete Saloutos/The Stock Market; 26 (tr), The Granger Collection, New York; 27 (all), Peter Van Steen/HRW

Chapter Two 28–29 (all), © Flip Nicklin/Minden Pictures; 31 (tl), John Langford/HRW; 32 (tr), Sam Dudgeon/HRW; 34 (tr), Sam Dudgeon/HRW; 35 (tr), Mary Kate Denny/PhotoEdit; 36 (all), Archive Photos; 38 (t), John Langford/HRW; 39 (bl), John Langford/HRW; 41 (tr), Charles D. Winters; 43 (t), © Stephen Dalton/Photo Researchers, Inc.; 44 (tl), Matt Meadows/Photo Researchers, Inc.; 46 (all), Richard Megna/Fundamental Photographs; 47 (tr), Sam Dudgeon/HRW; 48 (bc), Sam Dudgeon/HRW; 49 (bl), Digital Image copyright © 2005 EyeWire ; 49 (br), John Langford/HRW; 50 (tr, tl), Digital Image copyright © 2005 EyeWire ; 50 (bc), Bob Daemmrich/HRW; 52 (bl), Richard Megna/Fundamental Photographs; 53 (br), Sam Dudgeon/HRW; 54 (tl), © Flip Nicklin/Minden Pictures; 54 (bc), Sam Dudgeon/HRW; 55 (tc), © Ross Harrison Koty/Getty Images; 55 (cl), Dick Luria/Photo Researchers, Inc.; 55 (br), John Langford/HRW; 59 (all), Victoria Smith/HRW

Chapter Three 60–61 (all), Matt Meadows/Peter Arnold, Inc.; 63 (tl), Charlie Winters/Photo Researchers, Inc.; 63 (tr), Richard Megna/Fundamental Photographs; 64 (t), © A.T. Willett/Getty Images; 65 (tr), © Detlev Van Ravenswaay/Photo Researchers, Inc.; 66 (bc), Sam Dudgeon/HRW; 66 (br, bl), John Langford/HRW; 67 (bcr), Hugh Turvey/Science Photo Library/Photo Researchers, Inc.; 67 (br), Blair Seitz/Photo Researchers, Inc.; 67 (bcl), Leonide Principe/Photo Researchers, Inc.; 67 (tc, tr), Sam Dudgeon/HRW; 69 (br), © Tony Mcconnell/Photo Researchers, Inc.; 69 (tl), © Najlah Feanny/CORBIS SABA; 70 (t), © Cameron Davidson/Getty Images; 71 (cr), © Sinclair Stammers/SPL/Photo Researchers, Inc.; 72 (br), © Michael English/Custom Medical Stock Photo; 73 (tr), Hugh Turvey/Science Photo Library/Photo Researchers, Inc.; 75 (br), © Darwin Dale/Photo Researchers, Inc.; 76 (bl), Sovfoto/Eastfoto; 77 (bl), Richard Megna/Fundamental Photographs; 79 (br), Ken Kay/Fundamental Photographs; 81 (cr), Ken Kay/Fundamental Photographs; 82 (br), Stephanie Morris/HRW; 83 (all), John Langford/HRW; 84 (tl), Image copyright ©1998 PhotoDisc, Inc.; 84 (tr), Renee Lynn/Davis/Lynn Images; 84 (bl), Robert Wolf/HRW; 655 (tl), Leonard Lessin/Peter Arnold, Inc.; 86 (br), Sam Dudgeon/HRW; 657 (t), Index Stock Photography, Inc.; 87 (cr), Peter Van Steen/HRW; 89 (tr), Sam Dudgeon/HRW; 660 (br), Matt Meadows/Peter Arnold, Inc.; 90 (tl), Image copyright © 2005 PhotoDisc, Inc.; 91 (cr), Charles D. Winters/Photo Researchers, Inc.; 91 (bcr), © Mark E. Gibson; 91 (br), Richard Megna/Fundamental Photographs; 94 (tl), Dr. E. R. Degginger; 94 (tr), courtesy of the Raytheon Company; 95 (cr), © Underwood & Underwood/CORBIS

Chapter Four 96–97 (all), Data courtesy Marc Imhoff of NASA GSFC and Christopher Elvidge of NOAA NGDC. Image by Craig Mayhew and Robert Simmon, NASA GSFC.; 98 (b), Yoav Levy/Phototake; 99 (tr), Stephanie Morris/HRW; 99 (bl, br), John Langford/HRW; 100 (tl), John Langford/HRW; 100 (tr), Richard Megna/Fundamental Photographs; 102 (tl, tr), Fundamental Photographs, New York; 106 (tl), © Digital Vision Ltd.; 106 (tr), Courtesy www.vischeck.com (program)/©Digital Vision Ltd. (frogs); 107 (tr), © Yoav Levy/Phototake; 111 (tr), Sam Dudgeon/HRW; 111 (bl), Don Mason/The Stock Market; 112 (all), Victoria Smith/HRW; 113 (cr), © Steve Dunwell/Getty Images; 117 (br), Sam Dudgeon/HRW; 118 (tl), Yoav Levy/Phototake; 118 (br), © Digital Vision Ltd.; 122 (tr), Digital Image copyright © 2005; 122 (tl), M. Spencer Green/AP/Wide World Photos; 123 (bc), Photo courtesy R.R. Jones, Hubble Deep field team, NASA; 123 (cr), NASA

Lab Book/Appendix "LabBook Header", "L", Corbis Images; "a", Letraset Phototone; "b", and "B", HRW; "o", and "k", images ©2006 PhotoDisc/HRW; 124 (br), Sam Dudgeon/HRW; 125 (all), Richard Megna/Fundamental Photographs; 127 (br), Sam Dudgeon/HRW; 128 (br), HRW Photo; 129 (r), Sam Dudgeon/HRW; 131 (br), Sam Dudgeon/HRW; 132 (br), Sam Dudgeon/HRW; 138 (br), Victoria Smith; 139 (br), Victoria Smith; 145 (tr), Peter Van Steen/HRW; 145 (br), Sam Dudgeon/HRW; 159 (tr), Sam Dudgeon/HRW

Credits